Estimating with Microsof
Unlocking the Power for Home Builders

Jay Christofferson

BuilderBooks™
National Association of Home Builders
1201 15th Street, NW
Washington, DC 20005-2800
(800) 223-2665
www.builderbooks.com

Estimating with Microsoft Excel, Second Edition
Jay Christofferson
ISBN 0-86718-549-X

© 2003 by BuilderBooks™
of the National Association of Home Builders
of the United States of America

All rights reserved. No part of this book may be reproduced or utilized in any form or by any means, electronic or mechanical, including photocopying and recording or by any information storage and retrieval system without permission in writing from the publisher.

Cover design by Tim Kaage
Printed in the United States of America

Cataloging-in-Publication Data
[Insert CIP Data—To Come]

Disclaimer

This publication is designed to provide accurate and authoritative information in regard to the subject matter covered. It is sold with the understanding that the publisher is not engaged in rendering legal, accounting, or other professional service. If legal advice or other expert assistance is required, the services of a competent professional person should be sought.

—From a Declaration of Principles jointly adopted by a Committee of the American Bar Association and a Committee of Publishers and Associations.

For further information, please contact:
    BuilderBooks™
    National Association of Home Builders
    1201 15th Street, NW
    Washington, DC 20005-2800
    (800) 223-2665
    Check us out online at: www.builderbooks.com

01/03 Circle Graphics / Victor Graphics, Inc. 5000

About the Author

Jay Christofferson serves as chair of the Construction Management program at Brigham Young University. He received a Ph.D. in Construction Management from Colorado State University. Jay is a licensed general contractor and has built hundreds of custom and production homes and has been a consultant for many residential builders. He is the co-founder of Graduate Builder Seminars (GBS). He is also a frequent invited speaker at the NAHB International Builders' Show.

Through his commitment to developing computer solutions in management, communication, and estimating for construction companies, Jay developed an Excel-based estimating program for small to medium size builders and remodelers called Estimator*PRO*.

Acknowledgments

I would like to thank all those who have helped in any way and have given important input into the final product of the first edition and this book: to Kurt Lindblom from NAHB who worked with me from the inception of the idea and has given important direction and encouragement, and to Erica Orloff who has given much sound advice and has assisted with the presentation and flow of the material. I express appreciation to the faculty members and students at the Construction Management Program at Brigham Young University for their input into the technical details of and methods used in this book.

Special thanks to my wife, Maxine, and to my family, Michael, David, Janae, Kelli, and Ryan for their great support during the writing of this book.

Thanks also to Craig Weston for his technical input, to John Jones for his graphic support, and to Koreen Hansen for her close review of the manuscript.

Book Production

Estimating With Microsoft Excel, Second Edition was produced under the general direction of Gerald Howard, NAHB Executive Vice President and CEO, in association with NAHB staff members Michael Shibley, Executive Vice President, Builder, Associate, and Affiliate Services; Greg French, Staff Vice President, Publications and Non-dues Revenues; Eric Johnson, Publisher, BuilderBooks; Theresa Minch, Executive Editor; and Jessica Poppe, Assistant Editor.

Contents

Introduction .... **xix**

CHAPTER 1

Spreadsheet Basics and Not So Basics **1**
Getting Started 1
Using Workbooks 4
Naming Cells and Cell Ranges 8

CHAPTER 2

Overview of a Computerized Spreadsheet Estimate **19**
Cost Breakdown Summary Sheets 19
Detail Sheets 22
Databases 26

CHAPTER 3

Setting Up a Cost Breakdown Summary Sheet **29**
Using Cost Breakdown Summary Sheets 29
Writing Formulas 31
Copying Formulas to Other Cells 33
Paste Special 38
Sorting Data 42

CHAPTER 4

Creating Detail Sheets **45**
Detail Sheet Format 45
VLOOKUP 47
Data Validation 53

CHAPTER 5

Linking **59**
The IF Function 60
Linking 63
Hyperlink 65

CHAPTER 6
More on Formulas and Functions 71
Estimating Concrete 72
Calculating Rebar for Footings and Foundations 83
Estimating Permit Fees 86
Connection and Impact Fees 91
Roofing 94

CHAPTER 7
Loan Interest and Builder's Margin 103
Construction Loan Interest 103
Builder's Margin: Profit and Company Overhead 107
Circular References 110

CHAPTER 8
Automating Spreadsheets with Macros 113
Macros 114
Command Buttons 119
Attaching Macros to Objects 122
Creating Custom Icons 123
Editing Macros in Visual Basic for Applications 132

CHAPTER 9
Using Form Tools to Enhance Your Spreadsheets 135
Option Buttons 136
Check Boxes 140
Scroll Bar 142
List Boxes 144
Combo Boxes 147

CHAPTER 10
UserForms to Enter Data 149
Data Validation 149
UserForm 151
List Boxes 154

CHAPTER 11
Integration: Using Excel with Other Programs 165
Other Applications 165
Exporting Data 170
Object Linking and Embedding 174

Appendix: How to Use the CD 185

Resources 187

Index 189

Figures

Figure 1.1	Undo and Redo Buttons	4
Figure 1.2	Excel Workbooks	4
Figure 1.3	Insert Dialog Box	6
Figure 1.4	Setting the Default Cursor Movement	7
Figure 1.5	Selecting and Naming a Range of Non-Contiguous Cells	8
Figure 1.6	Creating a Cell Name Automatically	9
Figure 1.7	The Named Range Principal	9
Figure 1.8	Naming a Formula	10
Figure 1.9	Creating a Named Cell Reference	11
Figure 1.10	Pasting a Formula Name into a Cell	11
Figure 1.11	Modifying a Named Formula	12
Figure 1.12	Defining a Tax Constant	13
Figure 1.13	Using a Constant in a Formula	14
Figure 1.14	Naming Cell Ranges in a Table	15
Figure 1.15	Formula to Return the Value of Intersecting Ranges	16
Figure 1.16	Using AutoFill to Create Data Series	16
Figure 1.17	The Formatting Toolbar	17
Figure 1.18	Automatically Resizing Column Widths	18
Figure 2.1	Work Breakdown Items with Estimated Costs and Cost Variances	21
Figure 2.2	Cost Control Showing Negative Cost Variance	21
Figure 2.3	Roofing Cost Breakdown Item Estimate	22
Figure 2.4	Roofing Detail Estimate	23
Figure 2.5	Total Roofing Cost for 6/12 Sloped Roof	24
Figure 2.6	Roofing Detail for 6/12 Sloped Roof	25
Figure 2.7	Basic Format for Laying Out Detail Sheets	26
Figure 2.8	Database for Roofing Material	27
Figure 3.1	Sample Cost Breakdown Summary Sheet	30
Figure 3.2	AutoSum Button	32
Figure 3.3	Selecting F8:I8 in the Formula Bar	33

Figure 3.4	Using the AutoFill Feature	34
Figure 3.5	To Copy Formulas, Drag the Fill Handle Down	34
Figure 3.6	Using AutoSum to Sum Values in Columns	35
Figure 3.7	Function Wizard	35
Figure 3.8	Paste Function Dialog Box	35
Figure 3.9	Serial Number Representing the Current Date	36
Figure 3.10	Format Cells Dialog Box	37
Figure 3.11	Cell K3 Formatted to Current Date	37
Figure 3.12	Formatting Cells to Display Currency	38
Figure 3.13	Increase Decimal and Decrease Decimal Buttons	38
Figure 3.14	The Copy Button	39
Figure 3.15	The Paste Special Dialog Box	39
Figure 3.16	Paste Special Operations	41
Figure 3.17	Paste Special Multiply Operation	41
Figure 3.18	Paste Special, Transpose	42
Figure 3.19	Sorting Data	43
Figure 3.20	Data Sorted by State and then by City	44
Figure 4.1	General Form for Detail Sheets	46
Figure 4.2	Detail Sheet with Price Extensions and Total	47
Figure 4.3	Simple Roofing Database	48
Figure 4.4	Naming the RoofingDB Range of Cells	48
Figure 4.5	Choosing VLOOKUP from the Function Wizard	49
Figure 4.6	The VLOOKUP Dialog Box	50
Figure 4.7	Correcting the Lookup_Value (Cell A3) to Match an Item in Column 1 of the Roofing Database	52
Figure 4.8	Data Validation Criteria	54
Figure 4.9	Entering a Data Validation List Source	55
Figure 4.10	Data Validation Drop-Down List	56
Figure 4.11	Using Data Validation Drop-Down Lists to Select Items	57
Figure 5.1	Without a Description in Column A, VLOOKUP Returns the #N/A Error Sign	60
Figure 5.2	Viewing the Formulas in a Worksheet	62
Figure 5.3	The Roofing Detail Sheet —Values View	62
Figure 5.4	Creating a Link	63
Figure 5.5	Paste Link Can Be Used to Create a Link	64
Figure 5.6	Inserting Hyperlinks	65
Figure 5.7	Edit Hyperlink Dialog Box	66
Figure 5.8	The Browse Excel Workbook Dialog Box	66
Figure 5.9	Cell Address E17 Named RoofingTotal	67

Figure 5.10	Linking to a Named Cell	**68**
Figure 5.11	Hyperlink to File Specifications.doc	**69**
Figure 5.12	Hyperlink to NAHB Website	**69**
Figure 6.1	Restricting Spreadsheet Input to Take-Off Quantities	**73**
Figure 6.2	Cubic Yard Formulas	**74**
Figure 6.3	Adding a Waste Factor to the Cubic Yard Formula	**74**
Figure 6.4	Using the ROUNDUP Function to Round Up to Quarter-Yard Increments	**76**
Figure 6.5	VLOOKUP Function	**77**
Figure 6.6	Extending the Total Costs	**77**
Figure 6.7	Controlling the Display of Error Signs	**78**
Figure 6.8	Dealing with Multiple Suppliers	**78**
Figure 6.9	Setting Up to Use the Match Function	**80**
Figure 6.10	Nesting a MATCH Function within a VLOOKUP Function	**81**
Figure 6.11	Creating a List Validation to Quickly Choose a Concrete Supplier	**82**
Figure 6.12	Choosing from a Data Validation List	**83**
Figure 6.13	Formulas for Footing Rebar	**84**
Figure 6.14	Custom Formats	**85**
Figure 6.15	Calculating Construction Valuation	**86**
Figure 6.16	UBC Building Permit Fee Structure	**87**
Figure 6.17	Calculating Permit and Associated Fees	**88**
Figure 6.18	Formula References Cells in the Permit Fee Table	**90**
Figure 6.19	Connection and Impact Fee Database	**91**
Figure 6.20	Connection and Impact Fee Detail	**92**
Figure 6.21	Adding Comments to a Worksheet	**93**
Figure 6.22	Roof Plan View	**95**
Figure 6.23	Conversion Factor for the Roof Slope	**95**
Figure 6.24	Slope Factor Formula	**96**
Figure 6.25	Formula to Calculate the Quantity of Shingles	**97**
Figure 6.26	Roofing Labor Database	**99**
Figure 6.27	Roofing Labor and Roofing Sub-Bid	**99**
Figure 6.28	25-Year Asphalt Shingle Sub-Bid	**100**
Figure 6.29	Cedar Shakes Sub-Bid	**100**
Figure 7.1	Construction Loan Interest and Fees	**104**
Figure 7.2	Typical Construction S-Curve	**105**
Figure 7.3	Averaging the Monthly Loan Amounts	**107**
Figure 7.4	Breakout of Direct Costs, Company Overhead Costs (G&A), and Profit	**108**
Figure 7.5	Formulas that Form Circular References	**110**
Figure 7.6	The Results of Formulas Using Circular References	**110**

Figure 7.7	Activating Iteration to Enable Calculation of Circular References	111
Figure 7.8	Using the Circular Reference Toolbar to Trace Errors	112
Figure 8.1	Recording a Macro	114
Figure 8.2	The Stop Recording Toolbar	115
Figure 8.3	Creating a Macro to Enter the Company Name and Address	116
Figure 8.4	Opening the Stop Recording Toolbar	117
Figure 8.5	The Macro Dialog Box	118
Figure 8.6	Results from Running the CompanyAddress Macro from Cell B3	118
Figure 8.7	The Command Button Icon on the Forms Toolbar	119
Figure 8.8	Assigning a Macro to an Object	120
Figure 8.9	The Shortcut Menu for Objects	121
Figure 8.10	Changing the Appearance of an Object	122
Figure 8.11	Assigning a Macro to an Object	123
Figure 8.12	Naming a New Toolbar	124
Figure 8.13	Creating a Custom Button	125
Figure 8.14	Changing the Image on a Button	126
Figure 8.15	Creating a Custom Button Image	127
Figure 8.16	Positioning Bar to Insert a Custom Menu Item	129
Figure 8.17	Custom Menu Item Inserted into the Tools Menu	130
Figure 8.18	Assigning a Macro to a Menu Item	131
Figure 8.19	Making a Workbook Visible	132
Figure 8.20	Editing Macros through the Visual Basic Programming Language Editor	133
Figure 9.1	The Forms Toolbar	136
Figure 9.2	Creating an Option Button	137
Figure 9.3	Formatting the Option Button Control	138
Figure 9.4	Using an IF-Then Statement with Option Buttons	139
Figure 9.5	Using a Group Box to Cluster Option Buttons	140
Figure 9.6	Using Check Boxes	140
Figure 9.7	Anchored References in Formulas—$F$2	142
Figure 9.8	Using Scroll Bars	143
Figure 9.9	Spinner Control	144
Figure 9.10	List Box	145
Figure 9.11	Format Control for List Box or Combo Box	145
Figure 9.12	Making a List Box Work	146
Figure 9.13	Selecting Items from a Combo Box List	147
Figure 9.14	After an Item is Selected from a Combo Box	147

Figure 10.1	Detail Sheet with VLOOKUP Formulas	150
Figure 10.2	UserForms can be used to Enter Item Descriptions	151
Figure 10.3	Automating Detail Sheets	151
Figure 10.4	The Visual Basic Editor	152
Figure 10.5	Creating a UserForm	153
Figure 10.6	Creating Code to Open a UserForm	154
Figure 10.7	Adding a List Box to a UserForm	155
Figure 10.8	Adding Command Buttons to a UserForm	156
Figure 10.9	Using RowSource to Fill the ListBox with Roofing Items	157
Figure 10.10	VB Code to Copy Item Description on ListBox to Worksheet	158
Figure 10.11	Closing a UserForm	160
Figure 10.12	Creating a Double Click Event	161
Figure 10.13	Completed Code for the UserForm	162

Figure 11.1	Comma-Delimited Text File	166
Figure 11.2	Tab-Delimited Text File	166
Figure 11.3	Importing Data, Step 1	167
Figure 11.4	Importing Data, Step 2	168
Figure 11.5	Importing Data, Step 3	169
Figure 11.6	Text File Imported into Excel	170
Figure 11.7	The Save As Dialog Box	171
Figure 11.8	Reducing the Application Window Size	172
Figure 11.9	Resize the Excel Window by Dragging the Edge of the Screen	172
Figure 11.10	Arranging Application Windows	173
Figure 11.11	Pasting an Excel Worksheet into an MS Word Document	174
Figure 11.12	Paste Special Dialog Box	175
Figure 11.13	Embedding an Excel Spreadsheet into a Word Document	176
Figure 11.14	In-Place Editing of an Embedded Excel Object in a Word Document	177
Figure 11.15	Sample MS Project Schedule	179
Figure 11.16	Pasting a Link from Project to Excel	180
Figure 11.17	Looking up Information	180
Figure 11.18	Linked Word Document	181
Figure 11.19	Making a Change in the Start Date of Construction	182
Figure 11.20	Changes in the Schedule are Automatically Reflected in Excel	182
Figure 11.21	Changes to the Schedule are Instantly Reflected in Word	182
Figure 11.22	Select a Different Activity	183
Figure 11.23	Verifying the Automatic Updates to the MS Word Document	183

Foreword

Computerized estimating has become the standard in the construction industry over the last 15 years. As companies have computerized, they first worked on their accounting operations. The next priority for most builders has been to computerize their estimating functions. Of all the estimating programs that are available, the most widely used program in residential construction estimating is a simple spreadsheet.

In this book, *Estimating with Microsoft® Excel*, Jay Christofferson gives you step-by-step instructions on how to create a powerful spreadsheet estimating program that meets the needs of home builders and remodelers.

Jay is one of the leading experts in computerized estimating for home builders. He is a builder and has also helped other builders all around the country develop better management systems for their operations.

Excel is one of the most powerful tools any home builder can use. For most builders it can be just as powerful as your vehicle or your cellular phone and can bring you the same return. Jay does just what the title implies, he shows you how to unlock the power of Excel. He explains how to make it simple enough for anyone to use and yet powerful enough for you to base important management decisions on the information you generate from it.

Jay's approach to estimating is so simple anyone can learn it and yet so well thought-out that you can estimate everything needed to build a house three different ways:

1. Estimating every stick and brick
2. Subcontractor bids
3. Unit price estimating for each phase

For most home builders computerized estimating accomplishes three major objectives:

1. It is much faster (many builders can complete an entire detailed estimate in 3 to 4 hours instead of 3 to 4 days).

2. It improves the accuracy of your estimates. Even though estimates are completed quicker, they are also much more accurate.
3. Your estimate is more organized. It flows from the beginning of construction to the end and it breaks down the separate work packages by vendor. This makes purchasing and cost control much easier.

This could be one of the most important books you have ever read. Enjoy it!

Leon Rogers
President of Construction
Management Associates

Introduction

In an effort to pin down the bottom-line costs of a project, home builders and remodelers have developed various methods of estimating. Detailed estimating is the most accurate approach, but it also takes the most time to complete. In a 1999 survey at the International Builders' Show (Christofferson), builders reported taking an average of 12.7 hours to complete a detailed estimate. A quicker, but less accurate way to estimate, uses the "guesstimate" method. With this approach, the estimator wets his index finger, holds it high in the air, and enters the amount that comes to mind. In reality, most builders use a combination of methods; some costs in an estimate come from a trade contractor's bid, some are treated as allowances, other costs are estimated using detailed methods, and still other costs are estimated based on linear foot or square foot measures.

Computers have taken much of the drudgery out of estimating, increasing the accuracy of estimates and decreasing the time needed to complete them. In the same survey, builders reported that by using computer programs to generate their estimates, the average time required to complete each estimate dropped to 4.9 hours (Christofferson, 1999). Yet many complained that the estimating programs they have tried are extremely complicated, take a long time to learn, are too rigid to customize, or are very expensive.

An increasing number of builders and remodelers, both commercial and residential are using computer spreadsheets to solve their estimating problems (*Engineering News-Record*, August 12, 2002). Most builders already own spreadsheet programs and some, unknowingly, don't realize that their spreadsheet programs came loaded on their computers when they bought them. The benefits of using computerized spreadsheets are:

- They are inexpensive.
- They are easy to use.
- They can be customized to your style of estimating.
- They are fast.
- They are powerful.

Using spreadsheet estimating allows estimators to customize their spreadsheets to do exactly what they need. Utilizing spreadsheets can reduce the time it takes to do an estimate to one or two hours.

While building houses during the infancy of the personal computer era, I often wished for an easy-to-use computer software that was powerful enough to solve many of the management problems faced while building. Computerized spreadsheets provided that power and ease, and programs continue to improve.

Several years ago, I was approached by a builder wanting to know if there was a way to create a customized estimating spreadsheet that would be powerful and effective. In the university-level Quantity Take-off and Estimating class that I teach, we began using spreadsheets to solve some of the estimating problems that we faced. Over the years, we have developed procedures and methods using computerized spreadsheets to quickly and automatically solve many of the estimating problems that challenge builders and remodelers. While teaching estimating seminars to builders and remodelers, I have seen increasing interest by builders and remodelers who want to be able to develop their own customized estimating spreadsheets.

This book focuses on teaching home builders and remodelers how to develop customized estimating programs using Microsoft Excel. You will not only learn the basics (and much more) of using computer spreadsheets, but will learn step by step how to create your own automated spreadsheet estimating program. The knowledge that you will gain can easily be applied to solve other office needs as well, so even if you are not an estimator, you will still gain a tremendous amount by reading this book.

Conventions in this Book

Keystroke sequences are **bolded**

Buttons, Formulas, and **Cell** names are set off in the text and multi-level menu commands are separated by a forward slash (/) between each menu selection.

TIP

When there is a note of interest with additional information, it is identified by this symbol.

Jay Christofferson, *Program Chair*
Brigham Young University
Construction Management

Spreadsheet Basics and Not So Basics

CHAPTER 1

IN THIS CHAPTER
- Use Microsoft Excel Help.
- Learn how to use workbooks.
- Name cells and cell ranges.
- Enter series of data and text.

Getting Started

Throughout years of teaching spreadsheet estimating and using spreadsheets in my work, I have learned how to apply many spreadsheet functions and features to construction and more particularly, estimating. I have used and taught a variety of spreadsheet software programs and, in my opinion, have found Microsoft Excel to be the most intuitive and powerful of the spreadsheet programs.

What I have found in teaching spreadsheet estimating seminars is that most people do not understand the capacity and power that computerized spreadsheets have to automate and simplify so many office tasks. Many people use computer spreadsheets much as they would paper spreadsheets. Some improve over the use of paper spreadsheets by setting up their computerized spreadsheets to add, subtract, multiply and divide data. Others have become more advanced users and are proficient with many of the features of electronic spreadsheets.

During the introduction of the estimating seminars I teach, I typically ask the participants to rate themselves as to their expertise in using computerized spreadsheets. Most rate themselves as having some knowledge of using spreadsheets, many consider themselves to be spreadsheet novices, while only a small percentage rate themselves as being experts. Many of those who rated themselves as being experts have approached me after the seminars and confessed that their knowledge was somewhat limited based on what they had just learned. They realized, in a sense, that they had been hand sawing their lumber when they could have been using a power saw.

The knowledge you will gain by reading this book and by doing the examples will be like buying that power saw. Your new knowledge will allow you to complete office tasks such as estimating in a fraction of the time you would normally require.

You may have a demanding schedule and wonder how you can possibly find time to take on another project. A cartoon I once saw puts this into perspective. In the cartoon, a medieval general is leading his army into battle. The army is prepared for battle and arrayed with shields held high and spears in hand. Next to the general, there is his aid and a salesman with a machine gun that he is trying to peddle. The general's aid is trying to catch the attention of the preoccupied general but is waived off. The caption conveys the general's impatience with the situation. It states, "No! I can't be bothered with new technology . . . we have a battle to fight!"

Are you investing time now for future success? As a builder, you know of the payback for the time you spend to sharpen a saw or to oil a tool. Don't win the battle and loose the war. You will win the war if your knowledge and ability to apply new technology and skills are competitively on the leading edge.

This chapter begins with some basics of Excel spreadsheet use. There are, however, ideas and concepts introduced that even the spreadsheet "experts" will want to learn about. If you have quite a bit of experience with spreadsheets, you may want to skim through this chapter and only spend time on the topics you are unfamiliar with.

The next few chapters discuss methods of setting up computerized estimating spreadsheets. You will learn functions and features that will make your estimating time faster and more accurate. Other chapters teach how to automate many Excel tasks and how to integrate with other software programs.

Step-by-step examples are given throughout the book. There is an accompanying computer disk that contains files and examples that will be useful to you as you follow and do the examples in the book. Some chapters have two companion files on disk; one gives a starting point as you begin a new chapter, and the other file is the completed tutorial as you finish the chapter. Use the latter file for reference as you work through the chapter and to compare your finished product.

What You Need to Have

To make the best use of this book, you will need to have a copy of Microsoft Excel 2000 or Excel 2002. Both versions of Excel will run with either the Windows 95 operating system or with Windows NT.

Your hardware will make a big difference how well Excel performs. A minimum of 32 megabytes of RAM memory is recommended. Anything above that would be preferred. Your computer's processor should be at least 133 megahertz. Most new processors run at speeds of 1.6 gigahertz or faster.

It is always a good idea to use the latest version or most recent upgrade of a software program. Using the latest version of software will allow you to apply the latest and most powerful features available. Upgrades from previous versions of Excel and competitive upgrades from other brand spreadsheets are very inexpensive. You will also save yourself a lot of unnecessary frustration if you upgrade your computer occasionally (every four to five years). By using up-to-date hardware, your software will run much faster and you will be able to work more effectively. You don't need the latest and most expensive computer system, but you should buy hardware that is near the top quality. This will save costs by extending the useful life of your computer.

Finding Help

For specific questions about any Excel function, button, or feature, use **Help** on the **Menu** toolbar.

> Click **Help** and choose **Contents** or **Index**.

Contents gives instruction on general topics. **Index** is a quick way to get the specific information you want. Type in the first letters of the word you have questions about, the word will be highlighted and you can then select the word and display its instructions.

Undoing Mistakes

Don't worry about trying something new. If it doesn't do what you expected it to, click the (left-pointing) **Undo** button on the **Standard** toolbar. The first time you click the **Undo** button, the last instruction you performed will be undone. The second time you click it, the second to the last instruction you performed will be undone and so forth. If you decide that you really didn't want to undo what you have undone, then click on the (right-pointing) **Redo** button.

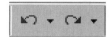

Figure 1.1 Undo and Redo Buttons

Using Workbooks

An Excel workbook is the file in which you store and analyze data. Figure 1.2 shows an Excel workbook containing three worksheets. The application name and the name of the open file (name of the workbook) are shown at the top. The **Menu** bar, **Standard** toolbar, and **Formatting** toolbar are shown near the top of the page. The **Drawing** toolbar has been placed near the bottom of the page.

The **Formula** toolbar is just above the column headings (A, B, C, etc.) and the **Status** toolbar is at the bottom of the page.

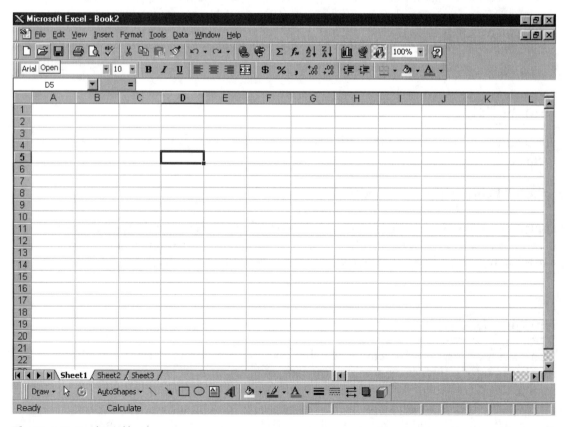

Figure 1.2 Excel Workbooks

> **TIP**
>
> To open or close toolbars, click **View/Toolbars** from the **Menu** bar and click on the toolbar that you want to open or close. An alternate, but faster way is to right mouse click over any one of the toolbars and then click on the toolbar name that you want to open or close.

Each worksheet is made up of columns (A through IV) and rows (1 through 65536). The intersection of a column and a row is called a cell. Cells are storage bins that can hold numbers, text, formulas, and formatting. Each cell has its own unique address, for example, the intersection of column D with row 5 is cell D5.

Adding and Deleting Worksheets

Similar to a folder in a file cabinet, the workbook contains any number of related sheets. Each worksheet takes up space and increases the size of the workbook, so unless it is needed for your application, you are better off with fewer sheets in your workbook. The default number of blank worksheets in a new workbook can be changed from 16 to some other number. This is accomplished through the **Tools** menu.

> Select **Tools/Options/General** and change the number of **Sheets in New Workbook** to the desired number of sheets that Excel will open in each new workbook.

Worksheets can be added to or deleted from a workbook at any time. To add a worksheet, click the right mouse button on one of the sheet tabs located at the bottom of the workbook. The new sheet will be inserted to the left of the sheet tab that you right-clicked.

To delete a worksheet, right click on the tab of the sheet that you want to delete then select **Delete** from the menu. A pop-up dialog box prompts you to confirm the deletion of the sheet or to cancel. Select **OK** to delete the sheet.

> **TIP**
>
> Click **Insert**, select the **General** tab (Figure 1.3) and select **Worksheet** (either double click the worksheet, or single click the worksheet and click **OK**).

Clicking the right mouse button (right click) will bring up related menus. Clicking the left mouse button (click) will select an item or a cell.

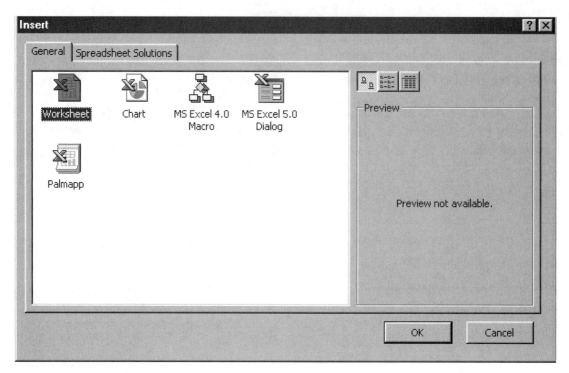

Figure 1.3 Insert Dialog Box

Naming a Worksheet

Worksheets can also be given names and re-ordered or moved.

- To name or rename a worksheet, double click on the tab of the sheet. The tab text is now in edit mode and can be changed.

Moving a Worksheet

- To move (re-order) a sheet, click on a tab and while holding the mouse button down, drag the tab to its new position.

Moving the Cursor

When data or text is typed and the **Enter** key is pressed, the cursor will automatically move down to the next cell. This default movement can be changed so that the cursor moves right, left, up, or does not move after the **Enter** key has been pressed.

To change the default movement of the cursor, click **Tools/Options/Edit** and make the desired changes (Figure 1.4).

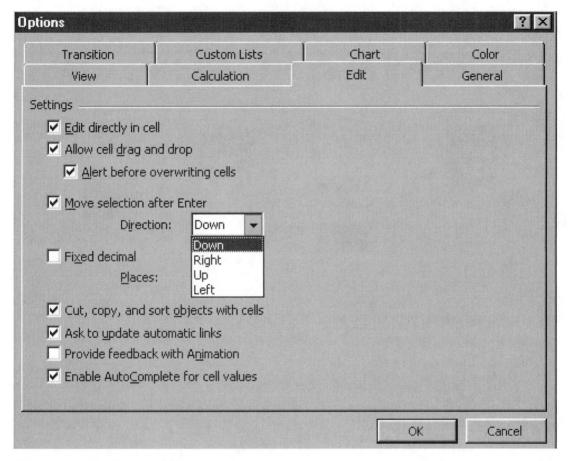

Figure 1.4 Setting the Default Cursor Movement

The cursor may also be controlled by selecting a range of cells and pressing the Enter key to move from cell to cell within the block.

Selecting a Block of Cells

To select a block of cells, select a cell in the corner of the block of cells and while holding the **Shift** key, click on a cell in the opposite corner of the desired block of cells. All the cells will be highlighted. As the **Enter** key is pressed, the cursor moves from cell to cell within the selected block of cells.

Selecting Non-Contiguous Cells

Non-contiguous cells, cells that are not together in a single block, can also be selected by holding down the **Ctrl** (control) key and selecting the cells you want (Figure 1.5).

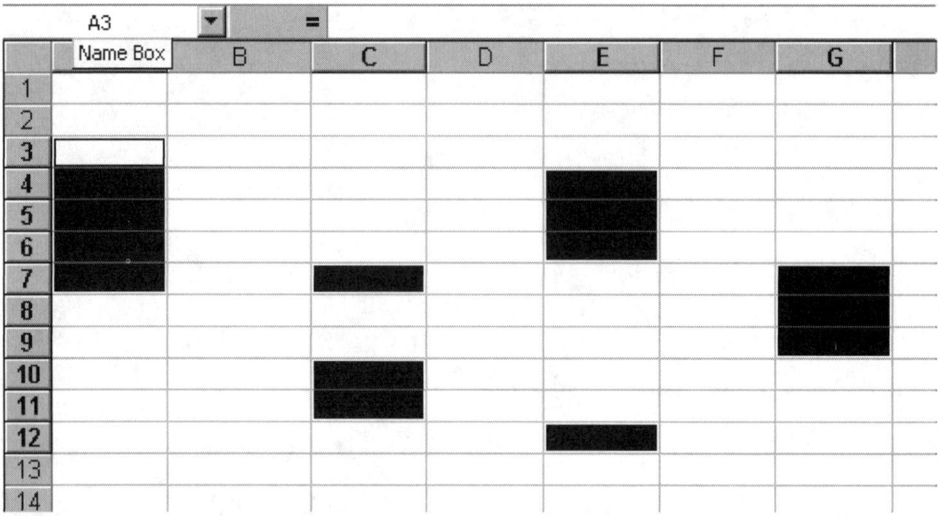

Figure 1.5 Selecting and Naming a Range of Non-Contiguous Cells

Pressing **Enter** will move the cursor from one cell to the next within the selected range of cells. Holding down the **Shift** key while pressing **Enter** causes the cursor to move in the opposite direction.

Naming Cells and Cell Ranges

Individual cells and ranges of cells can be named. The range of cells shown in Figure 1.5 was named **SelectedRange**.

To name a range, select the range of cells and type a one-word name into the **Name** box (on the left side of the **Edit** toolbar) then press **Enter**.

The named range, **SelectedRange**, can then be selected from any sheet in the workbook by selecting it from the names in the drop-down list in the Name box. The drop-down list is activated by clicking on the down arrow on the right of the **Name** box.

Cells that have either row or column headings can be named automatically.

In Figure 1.6, select both the row heading cells (**B3:B5**—read B3 through B5) and the cells to be named (**C3:C5**). Select **Insert/Name/Create**, then click **OK**.

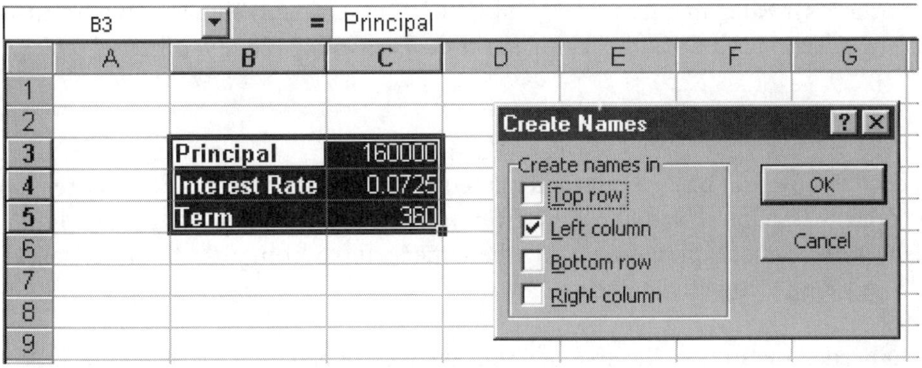

Figure 1.6 Creating a Cell Name Automatically

Cell **C3** is given the name **Principal**, cell **C4** gets **Interest_Rate**, and C5 is named **Term**. To check it out for yourself, click on cell **C3** and verify that the name **Principal** is shown in the **Name** box (see Figure 1.6).

Now, whenever the name Principal is selected from the drop-down list in the Name box, the cursor moves to C3 (Figure 1.7). Whenever the word Principal is used in a formula, the value in cell C3 will be substituted.

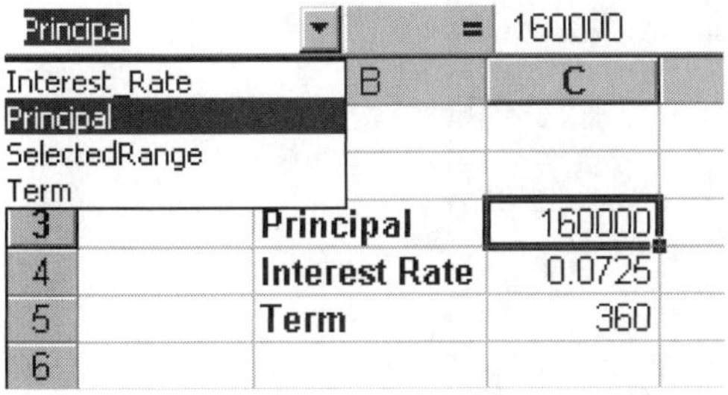

Figure 1.7 The Named Range Principal

For example, the following formula would return the interest for one month.

= Principal * Interest_Rate/12
($160,000 * .0725)/12 = $966.67

Formatting Cell for Currency and for Percentage

Cell **C3** (160000) can be formatted to currency style ($160,000.00) by selecting cell **C3** and clicking **$** on the **Formatting** menu toolbar. Cell **C4** (7.25) is correctly shown as a percentage (7.25%) by selecting cell **C4** and clicking **%** on the **Formatting** menu toolbar.

More on Names

You have seen the power of named cells and named cell ranges and there is even more that can be done with names. Names can be given to formulas that are used throughout the spreadsheet. Instead of typing or copying the same formula from place to place, name the formula and simply copy the name.

Suppose you needed to estimate the area of a round concrete pad for a gazebo. You could create a formula that calculates the area of a circle and that could be used anywhere in your workbook. The area of a circle is πr^2 and can be written this way:

=3.14*Radius^2

The Greek letter π (pi) represents a value of approximately 3.14. The letter r stands for the radius of the circle (Radius). The radius of the circle is squared (^2).

To create a name for this formula, click **Insert/Name/Define** and the **Define Name** dialog box will become visible (Figure 1.8). Enter the area of a circle formula in the **Refers to** box and the name **AreaCircle** in the **Names in workbook** box.

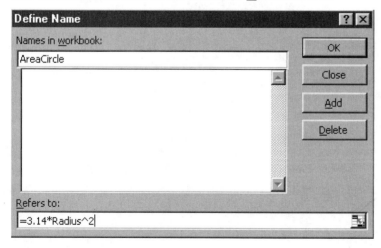

Figure 1.8 Naming a Formula

To use this formula in a worksheet, you would need to assign a value to the **Radius** variable. Enter the label **Radius** next to the cell where the value of the radius will be stored (cell **A1** and **B1**, Figure 1.9).

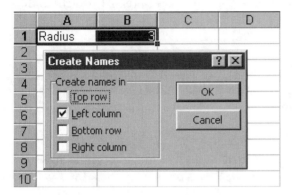

Figure 1.9 Creating a Named Cell Reference

Create the name for the variable (cell **B2**) by selecting cells **A1** and **B1** and then clicking **Insert/Name/Create**. The **Create Names** dialog box appears and asks if the name label is in the left column. Excel anticipates the correct answer and asks for verification. In our case, the name label is to the left of the cell were the radius' value will be stored. Make sure that a check mark is entered in the check box next to **Left column** in the dialog box, and then click **OK**.

Now paste the formula's name in the cell where you want the area of the circle to display (i.e. cell **D1**, Figure 1.10).

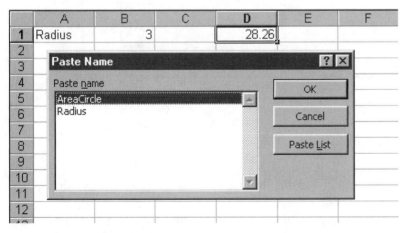

Figure 1.10 Pasting a Formula Name into a Cell

Click **Insert/Name/Paste** and select **AreaCircle** then click **OK**.

Figure 1.9 shows the result of the formula in cell **D1** and the formula is displayed in the Edit toolbar. To test the formula, change the value of the variable in cell **B1** and check the result.

Suppose you had used this formula in several places in the workbook and it became necessary to change the formula to get greater accuracy; use the more accurate 3.141593 instead of 3.14. Instead of having to change formulas in each of the several cells where the formula is used, because the formula is named, you simply have to change the formula in the named reference instead.

To change the formula for **AreaCircle**, click **Insert/Name/Define** and select **AreaCircle** (Figure 1.11). Change the value in the formula in the **Refers to:** box and click **OK**.

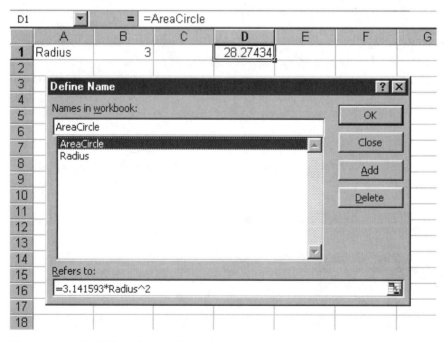

Figure 1.11 Modifying a Named Formula

All of the formula references will be automatically updated throughout the workbook.

Using named constants can be just as useful as using named formulas. The tax that is paid on the materials that you purchase must be accounted for when putting together an estimate. You don't estimate tax for all items, only for the materials that will be purchased. When you buy concrete, rebar, framing lumber, hardware, etc., you add a percentage (the tax rate) to the cost of the materials.

If you wanted to calculate the tax on concrete and your tax rate was 6.25 percent, you could write a formula that would take the subtotal of the concrete and multiply it by .0625. If the concrete subtotal was located in cell **G28**, the formula for tax could be located in **G29** and could be written.

=G28*.0625

You could write a formula similar to this one for every place in your estimate where tax is required. You might use the tax formula 15 or 20 times throughout your estimate. Then, suppose the tax rate increased to 6.375 percent. You would have to go back through each of the formulas and change the .0625 to .06375—a chore that would take a while to accomplish.

Figure 1.12 demonstrates the value of naming your constants. Instead of changing 15 or 20 tax formulas, you can use a constant (i.e. **Tax**) in the place of the tax rate (.0625). When the tax rate changes, the value of the constant can be changed and all references to the constant are automatically updated.

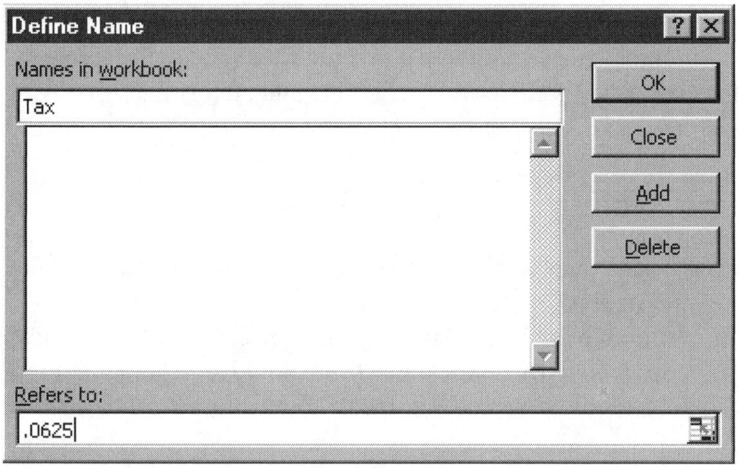

Figure 1.12 Defining a Tax Constant

To create a tax constant, click **Insert/Name/Define** on the **Menu** Bar. Enter the name of the constant (**Tax**) in the **Names in workbook:** box and enter the value of the constant (.0625) in the **Refers to:** box.

Figure 1.13 shows a how the **Tax** constant is used in the formula that calculates the tax amount:

=B1*Tax

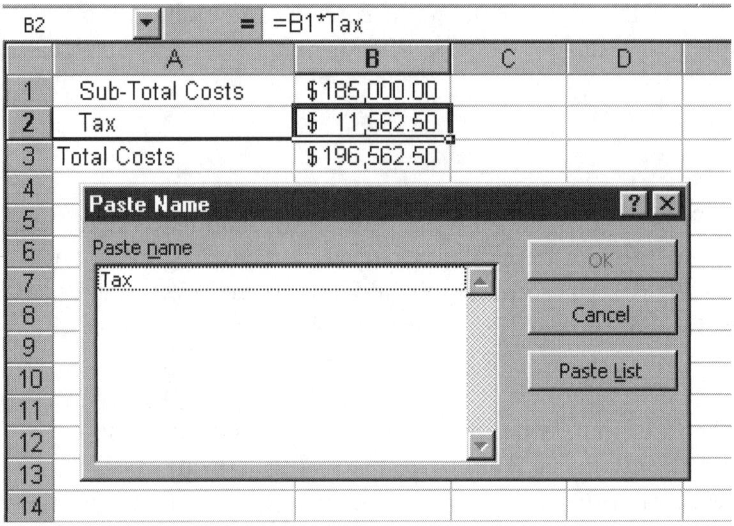

Figure 1.13 Using a Constant in a Formula

Cell B1 holds the cost of material to be taxed. When the tax rate changes, the tax constant can be changed in the **Define Name** dialog box (see Figure 1.11) and all formulas that use the Tax constant will be instantly updated.

Finding Information in a Table

Using named ranges can be especially helpful when searching for information in tables. In your business, you might be keeping track of the monthly sales for the different communities where you build. If you wanted to know the April sales for the Parkway community, and wanted to create a formula to look up this information, you could. By using the intersection operator with named ranges, you can create a formula to return specific data from a table.

First, name the rows and columns; you can do this in a single operation by selecting the table (include headers) and clicking **Insert/Name/Create** (Figure 1.13). The **Create Names** dialog box will prompt you to verify that the names for the columns are in the top row and that the names for the rows are in the first column. Click **OK** and all of the ranges will be named at once.

Now, wherever you may be in the workbook, a formula can be written that will instantly return the data that you need from the table. For example, you want to know what the April sales were for the Parkway community, enter the formula:

=April Parkway

The Intersection operator is the space between the two names. The value $749,000 is the intersection of the April range and the Parkway range (Figure 1.14). The formula, if rewritten, =Parkway April, would return the same value.

	A	B	C	D	E	F	G	H
1	Community	January	February	March	April	May	June	Community Total
2	Oak Hills	$ 345,000	$ 276,900	$ 498,000	$ 530,000	$ 428,700	$ 405,000	$ 2,483,600
3	Riverbrooke	$ 219,000	$ 450,000	$ 789,900	$ 810,240	$ 689,000	$ 655,900	$ 3,614,040
4	Parkway	$ 543,200	$ 437,000	$ 635,400	$ 749,000	$ 587,000	$ 567,900	$ 3,519,500
5	Meadow	$ 168,000	$ 356,000	$ 430,400	$ 560,700	$ 395,000	$ 405,700	$ 2,315,800
6	Sunwood	$ 688,200	$ 495,000	$ 578,100	$ 605,000	$ 525,800	$ 480,000	$ 3,372,100
7	Monthly Total	$1,963,400	$2,014,900	$2,931,800	$3,254,940	$2,625,500	$2,514,500	$15,305,040

Create Names
Create names in:
- ☑ Top row
- ☑ Left column
- ☐ Bottom row
- ☐ Right column

[OK] [Cancel]

Figure 1.14 Naming Cell Ranges in a Table

Entering Series of Data and Text

One of the more tedious tasks when setting up an estimate (or working with spreadsheets in general) is that of entering data and text. Figure 1.13 shows a table where the column headings are the months of the year. To type each of these months into the table could take from one to 10 minutes, depending on your typing skills. The **AutoFill** feature can eliminate most of the time needed for entering a series of data or text.

A series of numbers, dates, and other text can be created automatically using the **AutoFill** feature of Excel (Figure 1.15). After entering a value into a cell, the **Fill** handle (a small black square in the lower right corner of the cell) can be dragged to the desired location. When working with numbers, dragging the **Fill** handle will copy the number. Hold the **Ctrl** key while dragging and the cells will show an incremental change from the first cell.

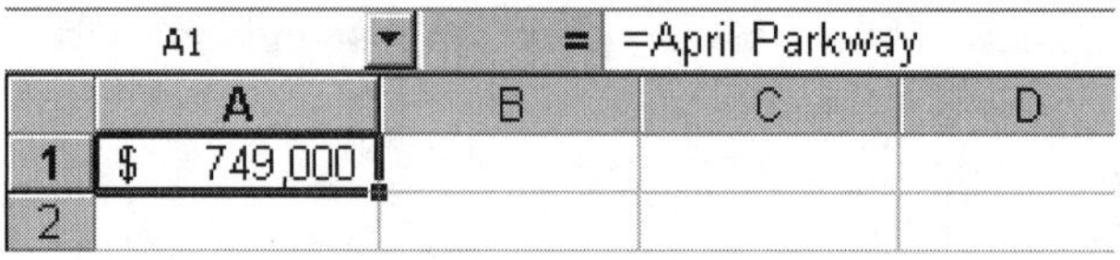

16   Estimating with Microsoft Excel

	A	B	C	D
	A1		=	=April Parkway
	A	B	C	D
1	$ 749,000			
2				

Figure 1.15 Formula to Return the Value of Intersecting Ranges

If you drag the **Fill** handle down or to the right, each cell will change in increasing order. By dragging the **Fill** handle up or to the left, the fill will be in decreasing order. To specify the amount of change (i.e. 0, 5, 10, 15, etc.), enter 0 in the first cell (A6, Figure 1.15), 5 in the second cell (B6), select both cells, then drag on the **Fill** handle (bottom right corner of the second cell) to the end of the series.

Formatting Cells

The values of the data and text on a spreadsheet are very important. The appearance of the data and text can be just as important. The way data and text appear on a spreadsheet can make a difference in its readability. Also important is the perception that others will have of your work as they read your spreadsheets. You should format your spreadsheets in such a way to establish an impression of professionalism.

The appearance, or formatting, of cells can be changed to fit specific needs. Text can be changed to different fonts or sizes. Text can also be bolded, italicized, colored, underlined, left, right, or center justified. Cell shading can be changed to any color and cells can be formatted for currency (C3, Figure 1.6), dates, percentage (C4, Figure 1.6), and many other standard and custom formats.

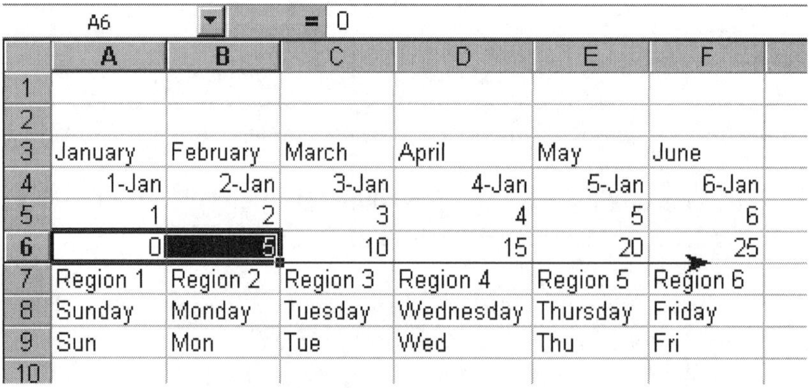

	A	B	C	D	E	F
	A6		=	0		
	A	B	C	D	E	F
1						
2						
3	January	February	March	April	May	June
4	1-Jan	2-Jan	3-Jan	4-Jan	5-Jan	6-Jan
5	1	2	3	4	5	6
6	0	5	10	15	20	25
7	Region 1	Region 2	Region 3	Region 4	Region 5	Region 6
8	Sunday	Monday	Tuesday	Wednesday	Thursday	Friday
9	Sun	Mon	Tue	Wed	Thu	Fri
10						

Figure 1.16 Using AutoFill to Create Data Series

The easiest and fastest way to format a cell is to select the cell and click the appropriate button on the Formatting toolbar (Figure 1.17).

Figure 1.17 The Formatting Toolbar

Listed from left to right, use the icons on the **Formatting** toolbar to change a cell's appearance or the appearance of the text in the cell:

Font Style	Align Left	Percentage Style	Increase Indent
Font Size	Center	Comma Style	Borders
Bold	Align Right	Increase Decimal	Fill Color
Italic	Merge Cells and Center	Decrease Decimal	Font color
Underline	Currency Style	Decrease Indent	

Other standard and custom formatting options are available by clicking Format from the Menu bar.

Moving and Copying the Contents of Cells

You will often find the need to move or copy information to other locations on your spreadsheet. The text, numbers, formulas, and formatting contained in cells can be moved or copied to other locations.

The contents of a cell (or cells) can be moved by selecting the cell (or cells) and then clicking **Edit/Cut** from the **Menu** bar. Select the new location and click **Edit/Paste**. An easier way to move a cell is to select it, move the cursor to the edge of the cell until the cursor turns to an arrow. Click on the edge of the cell and while pressing the left button, drag the cell to a new location and release the mouse button.

Copy the contents of the cell by selecting the cell (or cells) and by clicking **Edit/Copy** from the **Menu** bar. Select the new location and click **Edit/Paste**. An easier way to copy a cell is to select it, move the cursor to the edge of the cell until the cursor turns to an arrow. Hold the **Ctrl** key down. Notice that in **Copy** mode, a small + symbol appears next to the arrow cursor. Click on the edge of the cell and while pressing the left button and while pressing the **Ctrl** key, drag the cell to a new location and release the mouse button. A copy of the original cell (or cells) will be placed in the new location.

Changing the Height of Rows and Width of Columns

The standard width of columns may be too narrow or too wide for the data or text that are stored in them. The width of columns and the height of rows can be changed to fit the data and text stored in them.

If a column width needs to be changed, move the cursor to the right edge of the column heading. The cursor will change to a crosshair. Click and hold the mouse button down while dragging the column edge to the new size. Multiple columns can be re-sized to the same width by selecting across the column headings (click on a column label and drag across the desired number of columns), then dragging any one of the column heading edges to the desired width. All selected columns will change to the same width.

To quickly resize the width of a single column or multiple columns to fit the size of the data contained in the column or columns, select the column or columns and double click when the cursor changes to a crosshair at the right edge of the column heading (Figure 1.18). Each of the columns will automatically be resized to fit the largest text or data in each respective column.

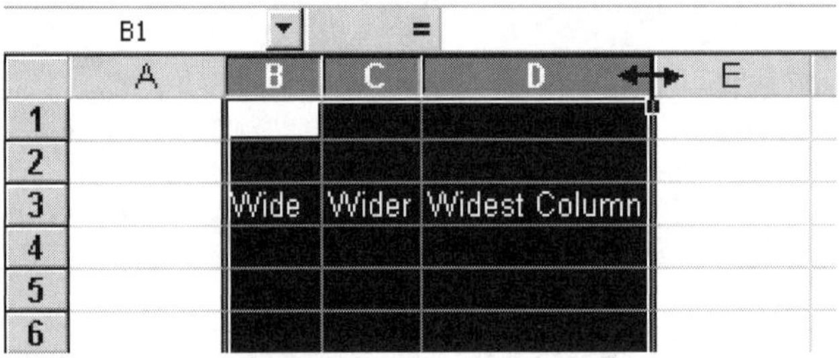

Figure 1.18 Automatically Resizing Column Widths

Row heights can be changed in a similar way. Select row headings instead of column headings and double click one of the border lines (where the cursor changes to a crosshair) between the row headings.

What's Next

If you've made it through this chapter, you will be further ahead in your Excel knowledge than the vast majority of contractors. What you have learned is important and will save you time as you begin to create solutions to solve office problems. Keep reading to learn about the real productivity tools found in the chapters that follow. Be sure to work through the examples as you read each chapter. You are on your way to unlocking secrets and the power of Excel.

Overview of a Computerized Spreadsheet Estimate

CHAPTER 2

IN THIS CHAPTER
- Create a custom estimating spreadsheet.
- Learn the basic format for a detail sheet.
- Develop a cost breakdown summary sheet.
- Use cost breakdown summary sheets for cost control.

Cost Breakdown Summary Sheets

If you were to look at your latest estimate, more than likely it would consist of a list of pre-construction and construction activities in their scheduled order or course of construction (for example, plans, permits, excavation, footings, foundation, etc.). Next to each activity would be an estimated cost. The estimate would probably be two or three pages long and would summarize all of the costs of construction.

This project Cost Breakdown Summary is used as a checklist to summarize and keep track of all the estimated costs of construction. Mortgage companies and mortgage departments of banks use cost breakdown summary sheets in conjunction with construction loans and paying draws against those loans. Builders, remodelers, and mortgage companies have all developed their own methods of organizing construction cost breakdowns.

In this chapter, you will begin to create a custom estimating spreadsheet. You will start by developing a cost breakdown summary sheet that will serve as a checksheet so that items are not overlooked in your estimate. This summary sheet will automatically retrieve summary costs from sheets containing the details of your cost estimating. The summary sheet can also serve as a tool to help keep track of and control costs.

Typical Cost Breakdown Summary Sheets

A typical cost breakdown summary may include the following direct costs:

Project Overhead

Lot	Temporary Utilities	Supervision
Permit and Fees	Construction Loan	Contingency
Plans & Engineering		

Hard Costs

Demolition	Windows	Floor Coverings
Earthwork	Plumbing	Cabinets
Footings	Heating	Countertops
Foundation	Air Conditioning	Appliances
Flatwork	Electrical	Hardware & Mirrors
Misc. Steel	Light Fixture Allowance	Siding
Window Wells	Roofing	Soffit & Fascia
Damp-proofing	Insulation	Gutter
Utility Laterals	Drywall	Exterior Railing
Septic System	Finish Carpentry Material	Foundation Plaster
Potable Water Well Allow	Finish Carpentry Labor	Clean-up
Framing Material	Painting	Landscaping
Framing Labor	Tile/Marble	Company Overhead (G&A)
Garage Doors	Fireplace Allowance	Builder's Margin

Some builders choose to follow Construction Specification Institute's (CSI) Master Format. In the CSI Master Format, construction costs are broken down into 16 major divisions and each major division is further subdivided into smaller divisions. Most commercial and industrial construction estimators use the CSI format to standardize the organization of estimates so that someone, other than the one doing the estimate, can easily find where costs are accounted for. CSI's sixteen major divisions are:

Division 1 General Requirements	Division 9 Finishes
Division 2 Site Work	Division 10 Specialties
Division 3 Concrete	Division 11 Equipment
Division 4 Masonry	Division 12 Furnishings
Division 5 Metals	Division 13 Special Construction
Division 6 Carpentry	Division 14 Conveying Systems
Division 7 Thermal & Moisture Protection	Division 15 Mechanical
Division 8 Doors & Windows	Division 16 Electrical

Construction cost breakdown summary sheets are typically two or three pages long. However, some builders have created detailed cost summary sheets that are several pages long. Each item on the cost breakdown summary sheet will have a cost budgeted to it. As costs are incurred against one of the work breakdown items, they are tallied and subtracted from the budgeted amount. The difference between estimated (budgeted) costs and the actual costs is called the cost variance. Tracking cost variances becomes the basis of cost control.

Using the Cost Breakdown Summary Sheet for Cost Control

Figure 2.1 shows a portion of a construction cost breakdown summary. The work breakdown items of demolition, earthwork, footings, and foundation are shown. The third column is where the estimated cost (budget) is entered. For footings, the estimated cost is $2,502.49. As invoices are received, they can be tallied in the appropriate cost category under one of the draw columns.

Code	Hard Costs	Est. Cost or Bid	Draw 1	Draw 2	Draw 3	Variance
	Demolition	$ -				$ -
	Earthwork	$ -				$ -
300	Footings	$ 2,502.49	$ 150.00	$ 2,100.00		$ 252.49
	Foundation	$ -				$ -

Figure 2.1 Work Breakdown Items with Estimated Costs and Cost Variances

In this example, the builder coded all invoices having to do with the footings with the number 300. In the first month of construction, $150.00 was paid to cover footing costs. In the next month, $2,100.00 in footing expenses was paid. The variance, $252.49, is the amount left in the footing budget to cover any additional costs. If no further footing expenses are incurred, this amount is profit left over at the end of the job.

The next month, the builder received a footing-related bill for $400.00 (Figure 2.2), the variance showed a negative $147.51. This represents the amount of money that is subtracted from the builder's profit margin. By knowing this early enough, the builder was able to make decisions to increase efficiencies in other areas during construction that resulted in having a profitable project.

Code	Hard Costs	Est. Cost or Bid	Draw 1	Draw 2	Draw 3	Draw 4	Total Draws	Variance
	Demolition	$ -					$ -	$ -
	Earthwork	$ -					$ -	$ -
300	Footings	$ 2,502.49	$ 150.00	$ 2,100.00	$ 400.00		$ 2,650.00	$ (147.51)
	Foundation	$ -					$ -	$ -

Figure 2.2 Cost Control Showing Negative Cost Variance

Because some builders and remodelers don't track variances during the course of construction, they do not know if they have made any money on a particular project until it is too late to do anything about it. By knowing that there are problems early in the project, builders can make a variety of adjustments to ensure a profit on the job. The builder could use less expensive suppliers or subcontractors, use alternate materials, or self-perform some of the work.

Detail Sheets

Each item on the cost breakdown summary has an estimated cost (budgeted amount) assigned to it. These estimated costs are derived using a combination of methods. Some budgeted amounts are determined by entering an allowance. Trade contractors provide bids for some of the cost breakdown items. Other amounts are calculated using price per square foot or price per linear foot. Many of the cost breakdown items are calculated using detailed quantity take-off and pricing methods. The detailed method is the most accurate way to estimate, but it requires the most time.

Figure 2.3 shows the roofing cost breakdown summary with an estimated amount of $2,906.86. The builder in this case bid the job 7 weeks before roofing was to begin. When asked what the cost would be to change from a 4/12-sloped roof to a 6/12-sloped roof, the builder could have guessed at the upgrade cost. However, to minimize the risk of guessing wrong, this builder had to completely redo the estimate because complete records of the details were not kept. Without keeping an itemized sheet containing the details of the roofing estimate, important information was forgotten and lost.

Electrical	$	-			$	-
Light Fixture Allowance	$	-			$	-
Roofing	$	2,906.86			$	2,906.86
Insulation	$	-			$	-
Drywall	$	-			$	-

Figure 2.3 Roofing Cost Breakdown Item Estimate

What-if Scenarios

When you look at a roofing figure like $2,906.86, you might ask what does this estimated amount include? Is this for a 4/12-, 5/12-, or 6/12-sloped roof? Does the amount include 30-year asphalt shingles or 30-year architectural asphalt shingles? Was ice and water shield included?

Figure 2.4 shows the detail that generated the roofing estimate amount in Figure 2.3. You can quickly see that among other items, 30-year asphalt shingles and two rolls of ice and water shield were included in the estimated cost for roofing. Roofing quantities were based on a 4/12-sloped roof.

Detail Material & Labor Description (Click Here)	Reset QTY	Unit	Choose: $/Unit	Supplier 1 Sub-Total $	Total Squares
Asphalt Shingles - 30 yr. 3 Tab	27.67	Sq (T)	$ 38.00	$ 1,051.33	27.67
Deliver & Stock shingles	30.33	Sq	$ 3.50	$ 106.17	
Drip Edge-metal	24.00	Ea	$ 3.90	$ 93.60	
30# Tar Paper (Asphalt-impregnated Felt)	14.00	2-Sq Roll	$ 10.95	$ 153.30	
Ice and Water Shield - 3' x 75'	2.00	2-Sq Roll	$ 77.00	$ 154.00	
Starter Strip	1.00	Sq (T)	$ 0.15	$ 0.15	1.00
Hip & Ridge Cap	1.67	Sq (T)	$ 0.32	$ 0.53	1.67
Turtle-back Vents 10 x 14 (1 SF net area)	4.00	Ea	$ 15.00	$ 60.00	
Step Flashing 3 x 4 x 7 (22' / Bndl.)	1.00	Bndl of 50	$ 12.95	$ 12.95	
Plastic Caps 1/2 lb./sq	2.00	Pail	$ 34.95	$ 69.90	
Roofing Nails - 1-1/4" - 2 lb./sq	56.00	Lb	$ 0.62	$ 34.72	
End					Total Squares
			Tax	$ 108.54	
			Material Subtotal:	$ 1,845.19	30.33

Detail Roofing Labor Labor Description (Type of Roof)	Slope	QTY	Unit	$/Unit	Sub-Total $
Asphalt Shingles - 30 yr. 3 Tab	4	27.67	Sq	$ 35.00	$ 968.33
Asphalt Shingles - 30 yr. 3 Tab	0	0.00	Sq	$ 35.00	$ -
Starter Strip, Hip, & Ridge	4	2.67	Sq	$ 35.00	$ 93.33
Total Squares:		30.33	Total Roofing Labor Cost:		$ 1,061.67

Figure 2.4 Roofing Detail Estimate

Computer spreadsheet detail sheets can instantly recalculate the costs of changing the roof slope from 4/12 to 6/12. Figure 2.5 shows the newly calculated total cost for a 6/12-sloped roof.

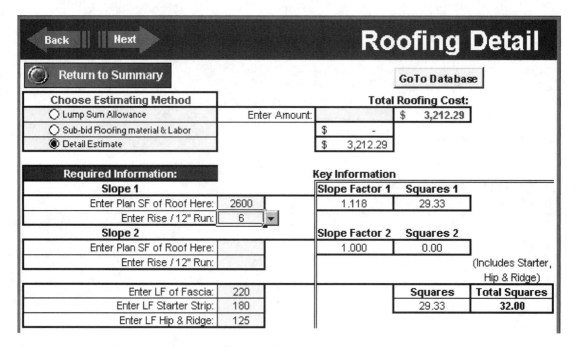

Figure 2.5 Total Roofing Cost for 6/12 Sloped Roof

Figure 2.6 shows the recalculated quantities for the 6/12-sloped roof. Now, instead of guessing or having to redo the roofing estimate, the builder could enter a rise of 6 into the spreadsheet, and instantly, the spreadsheet would recalculate the cost at $3,212.29. That's an increase of $305.43. There is no guessing, and the resulting answer is instantaneous.

Detail Material & Labor Description (Click Here)	Reset QTY	Unit	Choose: $/Unit	Supplier 1 Sub-Total $	Total Squares
Asphalt Shingles - 30 yr. 3 Tab	29.33	Sq (T)	$ 38.00	$ 1,114.67	29.33
Deliver & Stock shingles	32.00	Sq	$ 3.50	$ 112.00	
Drip Edge-metal	24.00	Ea	$ 3.90	$ 93.60	
30# Tar Paper (Asphalt-impregnated Felt)	15.00	2-Sq Roll	$ 10.95	$ 164.25	
Ice and Water Shield - 3' x 75'	2.00	2-Sq Roll	$ 77.00	$ 154.00	
Starter Strip	1.00	Sq (T)	$ 0.15	$ 0.15	1.00
Hip & Ridge Cap	1.67	Sq (T)	$ 0.32	$ 0.53	1.67
Turtle-back Vents 10 x 14 (1 SF net area)	4.00	Ea	$ 15.00	$ 60.00	
Step Flashing 3 x 4 x 7 (22' / Bndl.)	1.00	Bndl of 50	$ 12.95	$ 12.95	
Plastic Caps 1/2 lb./sq	2.00	Pail	$ 34.95	$ 69.90	
Roofing Nails - 1-1/4" - 2 lb./sq	59.00	Lb	$ 0.62	$ 36.58	
End					Total Squares
			Tax	$ 113.66	
			Material Subtotal:	$ 1,932.29	32.00

Detail Roofing Labor Labor Description (Type of Roof)	Slope	QTY	Unit	$/Unit	Sub-Total $
Asphalt Shingles - 30 yr. 3 Tab	6	29.33	Sq	$ 40.00	$ 1,173.33
Asphalt Shingles - 30 yr. 3 Tab	0	0.00	Sq	$ 35.00	$ -
Starter Strip, Hip, & Ridge	6	2.67	Sq	$ 40.00	$ 106.67
Total Squares:		32.00	Total Roofing Labor Cost:		$ 1,280.00

Figure 2.6 Roofing Detail for 6/12 Sloped Roof

Spreadsheets allow the user to evaluate what-if scenarios instantly and without effort. Questions dealing with costs such as, "What if we changed to 25-year asphalt shingles?" can be answered immediately. Computer spreadsheets make it simple, quick, and easy to calculate the costs associated with change orders.

Repair Shop Phenomenon

Using computer spreadsheets to calculate prices for upgrades and change orders has another advantage. I call it the repair shop phenomenon. If you take your car in for repairs, you may have some preconceived notions of how long it will take to repair the problem. You quickly multiply the time it should take by how much you think you should be paying per hour. But disappointingly, the manager goes to the computer and looks up the charges for your specific repair. No matter how much higher than your own estimate the actual price for the repair may be, you pay it without dickering. Who can argue with the computer? The price is what it is, and everyone gets charged the same amount.

By using computerized spreadsheets, your clients will have the same feeling. They will believe you are a professional with established costs, and they are less likely to question your figures.

Detail Sheet Format

The basic format for a detail sheet is shown in Figure 2.7.

	A	B	C	D	E
1					
2	Description	Quantity	Unit	$/Unit	Total $
3					
4					
5					
6					
7				Total Cost:	

Figure 2.7 Basic Format for Laying Out Detail Sheets

The formulas that calculate each work breakdown item's total cost, such as roofing, are entered into the cells of detail sheets. To generate item totals, quantities are multiplied by the cost per unit. The total cost is then calculated by summing all of the individual item totals.

Databases

Each detail sheet is supported by data such as units and unit prices that are stored in a database. A database is a list or a table of information. A roofing database may contain item descriptions, units of measure, unit prices, and other important information (Figure 2.8).

Overview of a Computerized Spreadsheet Estimate   27

	A	B	C	D
1		**Roofing Material Database**		
2		Description	Unit	$/Unit
3		Asphalt Shingles - 25 yr. 3 Tab	SQ	$ 30.00
4		Asphalt Shingles - 30 yr. 3 Tab	SQ	$ 38.00
5		Asphalt Shingles - 30 yr. Architectural	SQ	$ 48.43
6		Cedar Shakes #1 medium handsplits	SQ	$ 110.00
7		Cedar Shingles #1	SQ	$ 135.00
8		Metal Roofing	SQ	$ 90.00
9		Eagle Tile	SQ	$ 105.00
10		--		
11		Deliver & Stock shingles	SQ	$ 3.50
12		Drip Edge-metal	Ea	$ 3.90
13		15# Tar Paper (Asphalt-impregnated Felt	4-Sq Roll	$ 10.95
14		30# Tar Paper (Asphalt-impregnated Felt	2-Sq Roll	$ 10.95
15		Ice and Water Shield - 3' x 75'	2-Sq Roll	$ 77.00
16		Starter Strip	Sq (T)	$ 0.15
17		Hip & Ridge Cap	Sq (T)	$ 0.32
18		Architectural Hip & Ridge Cap	Box	$ 30.00
19		Ridge Vents	Lf	$ 2.30
20		Turtle-back Vents 10 x 14 (1 SF net area	Ea	$ 15.00
21		Step Flashing 3 x 4 x 7 (22' / Bndl.)	Bndl of 50	$ 12.95
22		18" x 10' Valley Flashing	Ea	$ 5.50
23		L Metal 4" x 4" x 10'	Ea	$ 4.79
24		Plastic Caps 1/2 lb./sq	Pail	$ 34.95
25		Simplex Nails-1/2 lb./sq	Lb	$ 0.85
26		Roofing Nails - 1-1/4" - 2 lb./sq	Lb	$ 0.62
27		Coil Roofing Nails 1" - 1-1/4"	Ea	$ 37.95
28		8d galvanized box nails	Lb	$ 0.64
29		6d galvanized box nails	Lb	$ 0.64
30		4d galvanized box nails	Lb	$ 0.64

Figure 2.8 Database for Roofing Material

Some builders still do much of their estimating by hand. When these builders create detailed estimates by hand, each line item description is written down and the quantity is calculated, usually by hand or with a handheld calculator. The unit or unit of measure is then entered next to the quantity. Next, the unit price has to be looked up and is written next to the unit of measure. The item total, the extended price, is the product of the quantity and the unit price. Some builders and estimators have preprinted forms that have a list of item descriptions, blank spaces where quantities can be entered, units of measure, and unit prices. To calculate the item totals, the quantities are multiplied by the corresponding unit prices using handheld calculators.

Typical Spreadsheet Estimating

When computerizing, preprinted hand forms are sometimes adapted to a spreadsheet so that the computer will automatically calculate the item totals. These item totals are then summed to find the page total. This method is far better than using hand forms, but a lot more can be done with computer spreadsheets to make the process faster and easier.

A Better Way to Use Spreadsheets to Estimate

Databases store information that can be accessed through formulas and functions that are entered into the detail spreadsheets. Informational items in the databases, such as item descriptions, units of measure, and unit prices, can be retrieved for use in the detail sheet.

Summary

We have talked about three types of computerized estimating spreadsheets:

1. The cost breakdown summary sheet, which is used to summarize the costs of all the work breakdown items of a project.
2. The detail sheets, which are used to list the specific items that will be part of the estimate (most of the calculations will be performed on the detail sheets).
3. The databases, which are used to provide information needed by the formulas in the detail sheets.

Start by Creating Your Own Cost Breakdown Summary Sheet

Chapter 3 provides the specific help you need to create your own customized Cost Breakdown Summary Sheet but first, begin by entering a list of all the work breakdown items that you would normally want to estimate. You can use the sample list of items presented earlier, the CSI divisions, or you can enter work breakdown items of your choice.

On the CD that accompanies this book is a file named **Chapter 2 Cost Breakdown Summary.xls**. It has a sample cost breakdown summary sheet that you can use as an example to create your own summary sheet. To open **Chapter 2 Cost Breakdown Summary.xls**:

1. Insert the CD into your computer's CD-ROM drive.
2. Open Excel and click **File/Open** then choose the CD-ROM drive letter.
3. Double click on the file name, or select the file and click **Open**.

CHAPTER 3

Setting Up a Cost Breakdown Summary Sheet

IN THIS CHAPTER

- Create standard and custom formulas.
- Use some of Excel's functions to make formulas.
- Create a cost breakdown summary sheet.
- Learn how to format cells.
- Sort data with more than one condition.

I worked with a builder who, in a year when sales were up, ended up making very little, if any, profit. All year long, decisions were made based on the premise that profit was being made on each job. Homes were estimated, sold, and built without ever comparing the actual costs with the estimated costs. It wasn't until tax time came along that the builder was even aware that the company had not made money that year. By then, it was too late to do anything about it.

If the builder had been able to compare actual costs with estimated costs as each home progressed, then he would have been aware of the losses early enough to correct most of them.

Using Cost Breakdown Summary Sheets

By using computerized spreadsheets to create and customize your cost breakdown summary sheets you will automate your work and save time. The cost breakdown summary sheet is a tool that can greatly simplify the estimating process and increase productivity for many reasons, including the following:

- It can instantly reflect any changes made in the details of the quantity takeoff sheets.

- It provides a professional-looking document that you can instantly print for mortgage companies, for buyers, if needed, or for your own office needs.
- The cost breakdown summary sheet also serves as a checklist so that items aren't overlooked in the estimate.
- It can also be used to control costs by tracking actual costs against budgeted costs (cost variances). This can make a great difference in the profitability of a builder or remodeler.

The layout of a cost breakdown summary sheet can take any form. The structure that has worked well for me is shown in Figure 3.1.

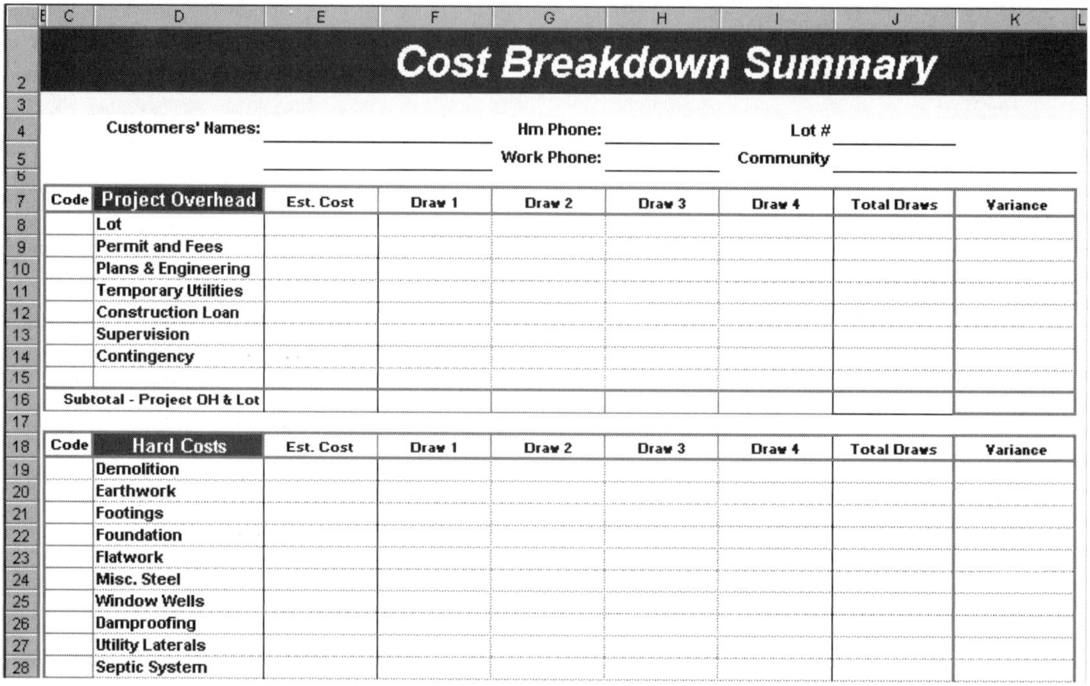

Cost Breakdown Summary

Customers' Names: ______ Hm Phone: ______ Lot # ______

______ Work Phone: ______ Community ______

Code	Project Overhead	Est. Cost	Draw 1	Draw 2	Draw 3	Draw 4	Total Draws	Variance
	Lot							
	Permit and Fees							
	Plans & Engineering							
	Temporary Utilities							
	Construction Loan							
	Supervision							
	Contingency							
Subtotal - Project OH & Lot								

Code	Hard Costs	Est. Cost	Draw 1	Draw 2	Draw 3	Draw 4	Total Draws	Variance
	Demolition							
	Earthwork							
	Footings							
	Foundation							
	Flatwork							
	Misc. Steel							
	Window Wells							
	Damproofing							
	Utility Laterals							
	Septic System							

Figure 3.1 Sample Cost Breakdown Summary Sheet

Layout of a Cost Breakdown Summary Sheet

This is the same summary sheet spoken of in Chapter 2 (**Chapter 2 Cost Breakdown Summary.xls**). The first column is for the cost code numbers. Cost code numbers can be any numbers that make sense to you. They can even be alphanumeric, a combination of alphabetic characters and numbers. You may want to use the numbers that are designated in the National Association of Home Builder's Chart of Accounts. NAHB has created an organized and standardized numbered list of accounts specifically for home builders and remodelers, available in *Accounting and Financial Management, Fourth Edition*, published by BuilderBooks.

As bills are received from suppliers and trade contractors during construction, the invoices are cost coded and the amounts of the invoices are entered into the appropriate draw (actual cost) columns. Cost variances are determined by subtracting the total of the draws from the estimated costs.

Writing Formulas

All formulas begin with the equal (=) sign. Cells are addressed storage boxes that can hold numbers. Numbers in cells can be changed, but the addresses of cells do not change. The data that is stored in cells is variable, in other words, it can be changed. Cell E8 for example, could hold the number 25,600 or 57,000, or any other number.

Formulas contain variables, littorals, and operators. Littorals are values in formulas that do not vary from use to use (i.e., 3, 26.7, and 100.65). Operators are commands that tell what to do with littorals or variables. A formula could be entered into cell E2, for example, to add the values of two variables. It might look like this:

=B6+C5

If the values stored in cells B6 and C5, respectively, were 4.5 and 72, the result of the formula would be 76.5. If, however, the variable in cell C5 were changed to 12, the result of the formula would be 16.5.

A formula that could be entered into cell D8, for example, and could add the values of two littorals might look like:

=3.14+56

The resulting value from this formula is 59.14. The following is a list of operators. The list gives an example how each operator is used and also, the resulting value:

Symbol	Operator	Example	Result
+	Addition	=2 + 5	7
−	Subtraction	=15 − 5	10
*	Multiplication	=4 * 3	12
/	Division	=20/10	2
^	Exponentiation	=3^2	3^2 or 9
%	Percentage	=170 * 15%	25.5

To find the variance costs, the estimated costs minus the total draws:

1. Click **K8**
2. Type =E8-J8:
    - Enter =
    - Click **E8**
    - Enter -
    - Click **J8**

Keep an eye on the **Formula** toolbar as you click on cells **E8** and **J8**. **E8** and **J8** are automatically entered into the formula.

The AutoSum Function

The first formula you may want to enter into the cost breakdown summary sheet might be a formula to calculate the total of all the actual costs for an item.

To sum the actual costs (draws):

1. Select cell **J8** in the **Total Draws** column (Figure 3.1).
2. Type the formula: =SUM(F8:I8).

To create this formula faster:

1. Select cell **J8**.
2. Click on the **AutoSum** button on the **Standard** toolbar (Figure 3.2). The formula =SUM() will automatically be entered into cell **J8**.

Figure 3.2 AutoSum Button

Setting Up a Cost Breakdown Summary Sheet

Now, select cells **F8** through **I8** (click on **F8** and while holding the left mouse button down, move to cell **I8**) and **F8:I8** will automatically be entered into the parenthesis.

=SUM(**F8:I8**)

Experiment with the summation formula. You will need to enter some numbers in cells F8 through I8.

To find out what the total of cells **F8** through **G8** are:

1. Select the text **F8:I8** in the **Edit** toolbar (Figure 3.3).
2. Select cells **F8** through **G8** (click **F8** and drag to **G8**).
3. Press **Enter**.

> **TIP**
>
> To calculate the total of F8, G8, and I8:
>
> 1. Select F8:G8 in the **Formula** toolbar.
> 2. Select cells F8 through G8 (click on F8 and drag to G8).
> 3. While pressing the **Ctrl** key, click on cell I8.
> 4. Press **Enter**.
>
> The formula =SUM(F8:G8,I8) will show in the formula bar and in cell J8. This is a powerful and timesaving feature.

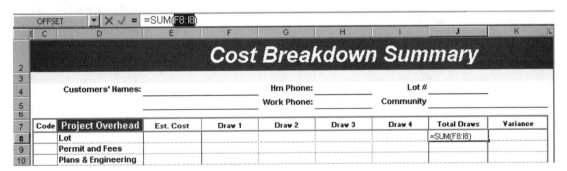

Figure 3.3 Selecting F8:I8 in the Formula Bar

Copying Formulas to Other Cells

You have created the formulas to total the draws and to calculate the variance between the estimated costs and the actual costs. You could repeat this procedure for rows 9 through 15 but there is an easier way to copy these formulas:

1. Select cells **J8** and **K8**.
2. Move the cursor to the **Fill** handle in the bottom right corner of the selected range (Figure 3.4). You will see the cursor change to a crosshair.

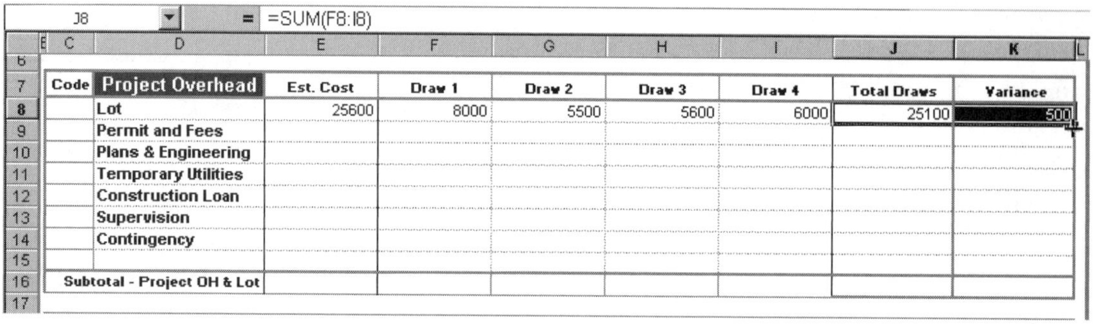

Figure 3.4 Using the AutoFill Feature.

3. Click on the crosshair and, while holding down the left mouse button, drag the **Fill** handle to **K15** (Figure 3.5).

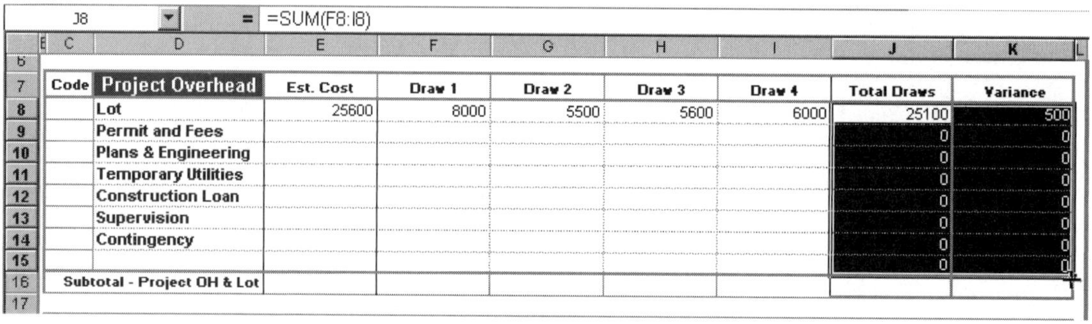

Figure 3.5 To Copy Formulas, Drag the Fill Handle Down

The formula in cell J8 will be copied to cells J9 through J15, and the formula in cell K8 will be copied to cells K9 through K15.

To find the total for the two columns, J and K:

1. Select cells **J16** and **K16**.
2. Click the **AutoSum** button. The summation formulas will automatically be entered into cells **J16** and **K16** (Figure 3.6).

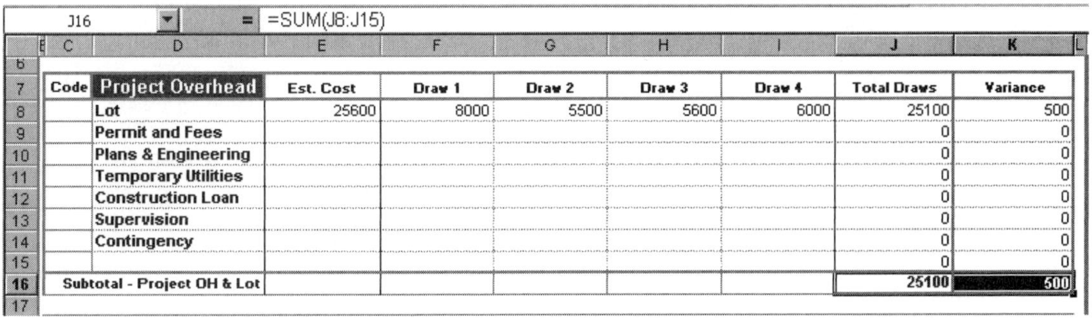

Code	Project Overhead	Est. Cost	Draw 1	Draw 2	Draw 3	Draw 4	Total Draws	Variance
	Lot	25600	8000	5500	5600	6000	25100	500
	Permit and Fees						0	0
	Plans & Engineering						0	0
	Temporary Utilities						0	0
	Construction Loan						0	0
	Supervision						0	0
	Contingency						0	0
							0	0
Subtotal - Project OH & Lot							25100	500

Figure 3.6 Using AutoSum to Sum Values in Columns

The Now Function

Another handy formula uses the **Now** function. You may want to show the current date whenever the cost breakdown summary sheet is opened. A quick way to enter the **Now** function into a formula is by using the **Function Wizard** on the **Standard** toolbar (Figure 3.7).

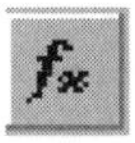

Figure 3.7 Function Wizard

1. Select cell **K3**
2. Click the **Function Wizard**. The **Paste Function** dialog box appears (Figure 3.8).

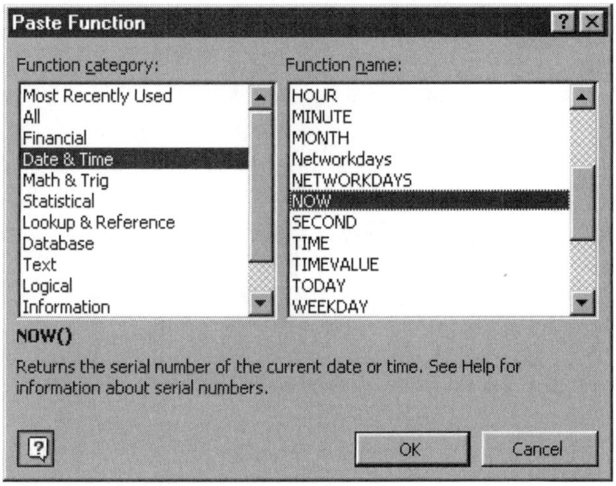

Figure 3.8 Paste Function Dialog Box

3. Select the function category **Date & Time**.
4. Select the function name **NOW**.
5. Click **OK**, or double click **NOW**.

Excel will insert the formula =NOW() into cell **K3**. If **K3** has not already been formatted, you will see the serial number that is the source code for the current date and time (Figure 3.9).

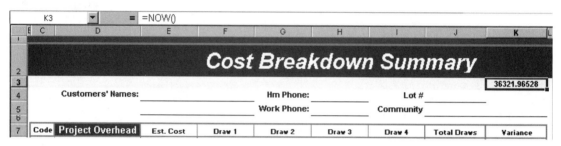

Figure 3.9 Serial Number Representing the Current Date

Formatting the Date

Date and time information can be formatted in many different styles.

> **TIP**
>
> Pressing **CTRL+;** will automatically enter the current date into the active cell. **CTRL+Shift+;** will automatically enter the current time.

1. Select cell **K3**.
2. Click **Format/Cells** and select the **Number** tab (Figure 3.10).
3. Select *either* the **Date** or the **Time** category.
4. Choose the configuration that you prefer from the predetermined formats and double-click.

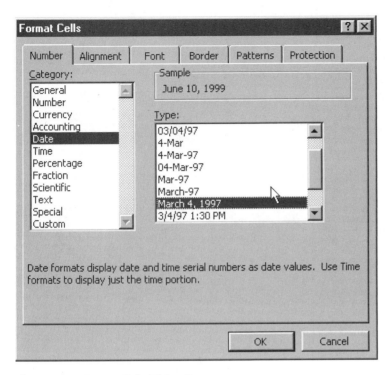

Figure 3.10 Format Cells Dialog Box

Your choice of the current date and/or time format will be displayed in cell K3 (Figure 3.11).

Figure 3.11 Cell K3 Formatted to Current Date

Formatting for Currency

Formatting the cells for currency will allow you to enter numbers that are unformatted, yet have Excel display the number with the dollar ($) sign in front and the appropriate comma (,) and period (.) placements. You may want to change the cells that hold currency data to the currency format.

1. Select the cells that you want to change.
2. Click the **Currency Style** button **($)** in the **Formatting** toolbar (Figure 3.12).

Figure 3.12 Formatting Cells to Display Currency

Increasing or Decreasing the Number of Zeros Behind the Decimal

You also can change the number of trailing zeros after the decimal:

1. Select the cells that you want to change.
2. Click on the **Increase Decimal** or **Decrease Decimal** buttons (Figure 3.13).

Figure 3.13 Increase Decimal and Decrease Decimal Buttons

It is better to set up your spreadsheet with all of the formulas before you do any formatting. When using **AutoFill** to copy formulas, the formats also are copied. You will end up redoing many of the formats.

Paste Special

One way to copy formulas without copying the formats also is to use the **Edit/Paste Special** command instead of AutoFill.

1. Select the cell that you want to copy.
2. Click **Edit/Copy** or simply click the **Copy** button on the **Menu** bar (Figure 3.14).

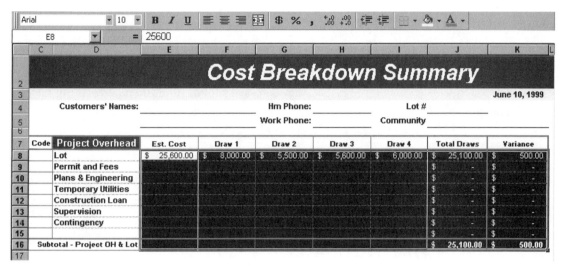

Figure 3.14 The Copy Button

Next, select the range of cells where you want to copy the formula to.

1. Click **Edit/Paste Specia**l. The **Paste Special** dialog box appears (Figure 3.15).
2. Click **Formulas**
3. Click **OK** or press **Enter**.

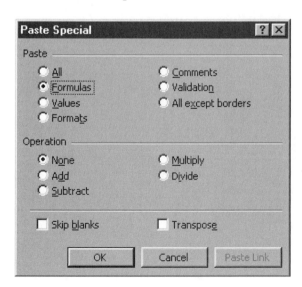

Figure 3.15 The Paste Special Dialog Box

The Paste Special Operators

Paste Special is worth spending some time to learn. It contains a few features that can be very helpful at times. One special group in the Paste Special dialog box contains the Operator functions (see Figure 3.15). They can save a lot of time when performing calculations. Let's look at how you can use this function in the following example.

Example

Your roofing supplier sends a notice that all roofing material prices will be increased 6 percent for the coming year. Recalculate all your figures? Maybe that used to be a nightmare, but with Excel you can use the Multiply operator of Paste Special to quickly update the costs.

Figure 3.16 shows unit costs located in column C. The new costs need to be entered into column D. The new prices will be 6 percent higher than the old prices.

To quickly add 6 percent to all of the current costs:

1. Enter 1.06 into **D3** and copy down to **D19**.
2. Select cells **C3:C19** and click the **Copy** button.
3. Select **D3**.
4. Click **Edit**.
5. Click **Paste Special**.
6. Select the Operator, **Multiply**.
7. Click **OK** or press **Enter**.

The result of the Multiply operation is that the product of Column C and Column D is pasted into Column D (Figure 3.17).

Setting Up a Cost Breakdown Summary Sheet   41

	A	B	C	D
1	**Roofing Material Database**			
2	Description	Unit	Old $/Unit	New $/Unit
3	Asphalt Shingles - 25 yr. 3 Tab	SQ	$ 30.00	$ 1.06
4	Asphalt Shingles - 30 yr. 3 Tab	SQ	$ 38.00	$ 1.06
5	Asphalt Shingles - 30 yr. Architectural	SQ	$ 48.43	$ 1.06
6	Cedar Shakes #1 medium handsplits	SQ	$ 110.00	$ 1.06
7	Cedar Shingles #1	SQ	$ 135.00	$ 1.06
8	Metal Roofing	SQ	$ 90.00	$ 1.06
9	Eagle Tile	SQ	$ 105.00	$ 1.06
10	--------------------------------------			
11	Deliver & Stock shingles	SQ	$ 3.50	$ 1.06
12	Drip Edge-metal	Ea	$ 3.90	$ 1.06
13	15# Tar Paper (Asphalt-impregnated Felt)	4-Sq Roll	$ 10.95	$ 1.06
14	30# Tar Paper (Asphalt-impregnated Felt)	2-Sq Roll	$ 10.95	$ 1.06
15	Ice and Water Shield - 3' x 75'	2-Sq Roll	$ 77.00	$ 1.06
16	Starter Strip	SQ	$ 0.15	$ 1.06
17	Hip & Ridge Cap	SQ	$ 0.32	$ 1.06
18	Architectural Hip & Ridge Cap	Box	$ 30.00	$ 1.06
19	Ridge Vents	LF	$ 2.30	$ 1.06

Figure 3.16 Paste Special Operations

	A	B	C	D
1	**Roofing Material Database**			
2	Description	Unit	Old $/Unit	New $/Unit
3	Asphalt Shingles - 25 yr. 3 Tab	SQ	$ 30.00	$ 31.80
4	Asphalt Shingles - 30 yr. 3 Tab	SQ	$ 38.00	$ 40.28
5	Asphalt Shingles - 30 yr. Architectural	SQ	$ 48.43	$ 51.34
6	Cedar Shakes #1 medium handsplits	SQ	$ 110.00	$ 116.60
7	Cedar Shingles #1	SQ	$ 135.00	$ 143.10
8	Metal Roofing	SQ	$ 90.00	$ 95.40
9	Eagle Tile	SQ	$ 105.00	$ 111.30
10	--------------------------------------			
11	Deliver & Stock shingles	SQ	$ 3.50	$ 3.71
12	Drip Edge-metal	Ea	$ 3.90	$ 4.13
13	15# Tar Paper (Asphalt-impregnated Felt)	4-Sq Roll	$ 10.95	$ 11.61
14	30# Tar Paper (Asphalt-impregnated Felt)	2-Sq Roll	$ 10.95	$ 11.61
15	Ice and Water Shield - 3' x 75'	2-Sq Roll	$ 77.00	$ 81.62
16	Starter Strip	SQ	$ 0.15	$ 0.16
17	Hip & Ridge Cap	SQ	$ 0.32	$ 0.34
18	Architectural Hip & Ridge Cap	Box	$ 30.00	$ 31.80
19	Ridge Vents	LF	$ 2.30	$ 2.44

Figure 3.17 Paste Special Multiply Operation

Paste Special/Transpose

The **Transpose** function of **Paste Special** is very useful at times. You may have a list of names or numbers that you want transferred from a vertical column to a horizontal row or vice versa. In Figure 3.18, for example, you may want to change the vertically placed city names (A3:A13) to arrange them horizontally.

	A	B	C	D	E	F	G	H	I	J	K
1	Atlanta	Dallas	Seattle	St. Louis	Richmond	San Diego	Buffalo	Orlando	St. Paul	Boise	Tulsa
2											
3	Atlanta										
4	Dallas										
5	Seattle										
6	St. Louis										
7	Richmond										
8	San Diego										
9	Buffalo										
10	Orlando										
11	St. Paul										
12	Boise										
13	Tulsa										

Figure 3.18 Paste Special, Transpose

To transpose data:

1. Copy the cells that contain the list of names.
2. Select the starting cell of the horizontal segment (A1 in this case)
3. Click **Edit**.
4. Click **Paste Special**.
5. Select **Transpose**.
6. Click **OK**.

Sorting Data

As you look at the list of cities, it may feel unnatural to you to see the list out of alphabetical order. Excel has a powerful function to sort data. In the example (Figure 3.19), the cities are located next to their corresponding states. To sort this list alphabetically by state:

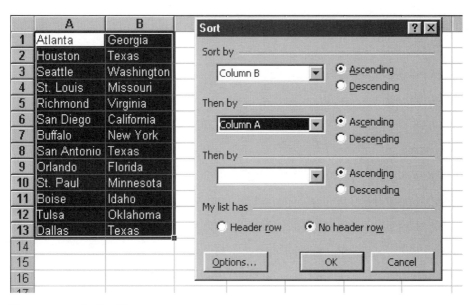

Figure 3.19 Sorting Data

1. Select the list of cities and states.
2. On the **Menu** toolbar click **Data**.
3. Click **Sort**.

The **Sort** dialog box appears. You are asked which column you would like to sort by, either by cities (Column A), or by states (Column B). You can sort in Ascending order (alphabetically) or Descending order (reverse alphabetic order).

In cases like this where there are many cities for each state, it may be necessary to sort on a second set of criteria, such as:

1. Sort by state (Column B in ascending order).
2. Then sort by city (Column A in ascending order).
3. In this case, there is no header row. Select the option button for **No Header Row** as shown in Figure 3.19.

Notice, as the data has been sorted (Figure 3.20), that the second sort order criteria arranged the cities belonging to the same states in alphabetical order.

	A	B
1	San Diego	California
2	Orlando	Florida
3	Atlanta	Georgia
4	Boise	Idaho
5	St. Paul	Minnesota
6	St. Louis	Missouri
7	Buffalo	New York
8	Tulsa	Oklahoma
9	Dallas	Texas
10	Houston	Texas
11	San Antonio	Texas
12	Richmond	Virginia
13	Seattle	Washington

Figure 3.20 Data Sorted by State and Then by City

Summary

In this chapter, we discussed how to create a cost breakdown summary sheet. You learned how to create custom formulas and how to use some of Excel's functions to make formulas. Additionally, you were introduced to some key formatting options.

If you haven't already done so, take time now to complete your custom cost breakdown summary sheet. If you have problems with your own summary sheet, a sample summary sheet, **Chapter 3 Cost Breakdown Summary.xls** can be used to help you. The sample sheet is on the accompanying CD.

CHAPTER 4

Creating Detail Sheets

IN THIS CHAPTER

- Learn functions, formulas, and methods for spreadsheet estimating.
- Use VLOOKUP to coordinate price changes and calculations.
- Learn how to use the AutoSum button.
- Use Data Validation to streamline data entry.
- Create a roofing detail sheet and a roofing database.

Detail sheets are the workhorses of spreadsheet estimating. The formulas and functions that are used in creating detail sheets perform the calculations and do most of the timesaving work. A well-designed detail sheet can take most of the work out of estimating and significantly reduce the time to complete an estimate.

Detail Sheet Format

Detail sheets generally take the form shown in Figure 4.1. The estimator enters a description for each item; space is then allotted where an estimator can enter the quantity of each item. Units of measure are needed to describe the significance of the quantities that are assigned to each item; unit costs are also specified for each item. Of course, the detail sheet will do the cost extensions and total the extended costs. But detail sheets can do *much* more.

	A	B	C	D	E
1	**Roofing Detail**				
2	Description	Quantity	Unit	$/Unit	Total $
3	Shingles		SQ	$ 29.50	
4	Felt		Roll	$ 10.95	
5	Drip Edge		EA	$ 3.85	
6	Ice & Water Shield		Roll	$ 68.50	
7	Starter Strip		SQ	$ 29.50	
8	Hip & Ridge		SQ	$ 29.50	
9	Roofing Nails		LB	$ 0.78	
10	Total				
11					

Figure 4.1 General Form for Detail Sheets

Extending Item Costs

To calculate the item total costs (column E), you will want to create a formula to return the product of the Quantity and the Unit Price:

1. Click **E3**.
2. Type =B3*D3. Remember, after entering the = sign in cell **E3**, click on cell **B3** to automatically enter **B3** into the formula. Do the same for **D3** after entering the * sign.
3. Copy the formula to cells **E4:E9** by dragging **E3's** fill handle to cell **E9**.

To calculate the total of all the roofing items:

1. Select cell **E10**.
2. Click the **AutoSum button** (_).

You will need to enter some values into the Quantity column to check your formulas. The practice roofing detail sheet should look something like Figure 4.2.

	A	B	C	D	E
1	**Roofing Detail**				
2	Description	Quantity	Unit	$/Unit	Total $
3	Shingles	28	SQ	$ 29.50	$ 826.00
4	Felt	7	Roll	$ 10.95	$ 76.65
5	Drip Edge	18	EA	$ 3.85	$ 69.30
6	Ice & Water Shield	2	Roll	$ 68.50	$ 137.00
7	Starter Strip	1	SQ	$ 29.50	$ 29.50
8	Hip & Ridge	1.67	SQ	$ 29.50	$ 49.27
9	Roofing Nails	56	LB	$ 0.78	$ 43.68
10	Total				$ 1,231.40
11					

Figure 4.2 Detail Sheet with Price Extensions and Total

Extending the costs (QTY * Unit $) automates the process somewhat, but what about the cost per unit? Did you know what the unit cost was off the top of your head, or did you have to take some time and go to the file cabinet to get the current price? Even if you knew the unit cost, you still had to take the time to enter it into the $/unit column. The computer can do that for you automatically.

VLOOKUP

The **VLOOKUP** function can look up and return data from another worksheet or from another workbook. **VLOOKUP** is a function that may be a bit hard to grasp at first, but the concept is really rather simple. Once you understand it, **VLOOKUP** may become one of your most valuable tools when doing any kind of spreadsheet work.

We will start by creating a database that holds important information such as item descriptions, units of measure, and costs per unit. A simplified roofing database is shown in Figure 4.3. It was created on the same worksheet as the roofing detail to simplify the demonstration of **VLOOKUP**. Normally the database would be in another worksheet or in a separate workbook.

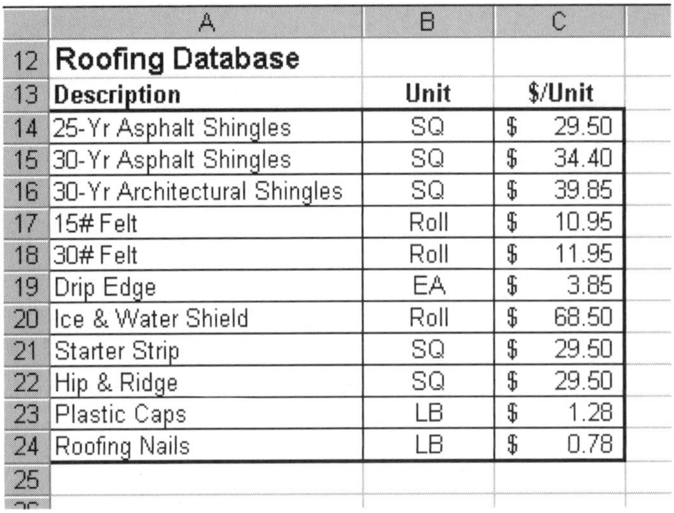

48 Estimating with Microsoft Excel

	A	B	C
12	**Roofing Database**		
13	Description	Unit	$/Unit
14	25-Yr Asphalt Shingles	SQ	$ 29.50
15	30-Yr Asphalt Shingles	SQ	$ 34.40
16	30-Yr Architectural Shingles	SQ	$ 39.85
17	15# Felt	Roll	$ 10.95
18	30# Felt	Roll	$ 11.95
19	Drip Edge	EA	$ 3.85
20	Ice & Water Shield	Roll	$ 68.50
21	Starter Strip	SQ	$ 29.50
22	Hip & Ridge	SQ	$ 29.50
23	Plastic Caps	LB	$ 1.28
24	Roofing Nails	LB	$ 0.78
25			

Figure 4.3 Simple Roofing Database

In Chapter 1, we talked about naming ranges of cells. To name the roofing database:

1. Select cells **A14:C24**.
2. In the **Name** box (upper left), enter RoofingDB (Figure 4.4).
3. Press **Enter**.

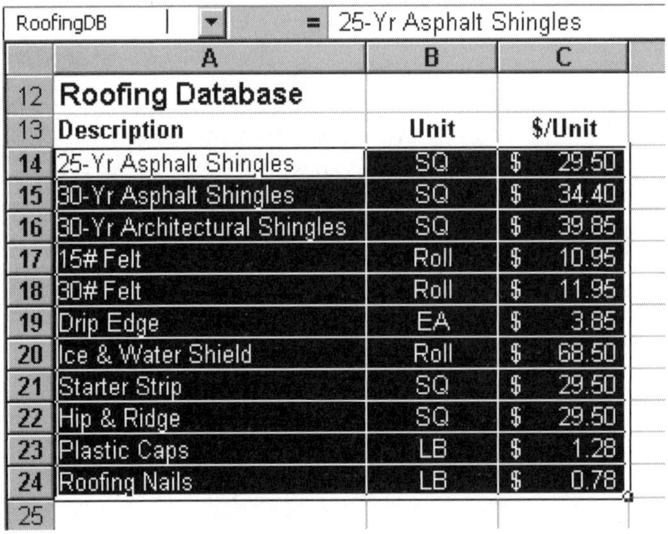

RoofingDB		=	25-Yr Asphalt Shingles	
	A	B	C	
12	**Roofing Database**			
13	Description	Unit	$/Unit	
14	25-Yr Asphalt Shingles	SQ	$ 29.50	
15	30-Yr Asphalt Shingles	SQ	$ 34.40	
16	30-Yr Architectural Shingles	SQ	$ 39.85	
17	15# Felt	Roll	$ 10.95	
18	30# Felt	Roll	$ 11.95	
19	Drip Edge	EA	$ 3.85	
20	Ice & Water Shield	Roll	$ 68.50	
21	Starter Strip	SQ	$ 29.50	
22	Hip & Ridge	SQ	$ 29.50	
23	Plastic Caps	LB	$ 1.28	
24	Roofing Nails	LB	$ 0.78	
25				

Figure 4.4 Naming the RoofingDB Range of Cells

4. Name the range of cells that include the list of descriptions for the roofing items.
5. Select cells **A14:A24**.
6. Type RoofingList in the **Name** box.
7. Be sure to press the **Enter** key to complete the naming process.

Notice that the roofing database (RoofingDB) has three columns:

- Column 1 of the roofing database (located in column A) holds the description data.
- Column 2 holds the units of measure for each of the roofing items.
- Column 3 stores the unit costs.

You are now ready to create the VLOOKUP formula that will find the unit price in the roofing database and enter it into the $/Unit column for the corresponding roofing item.

1. First, move to the roofing detail section.
2. Select cell **D3** in the **$/Unit** column (Figure 4.5).

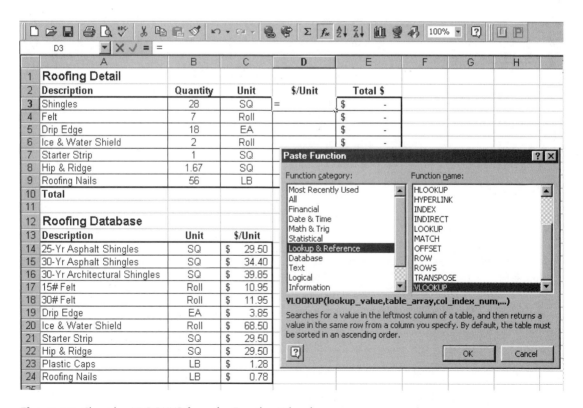

Figure 4.5 Choosing VLOOKUP from the Function Wizard

To aid you in the creation of the formula:

3. Click the **Function Wizard (fx)** on the **Standard** toolbar.
4. Click **Lookup & Reference** in the **Function** category.
5. Choose **VLOOKUP** from the **Function** name list.
6. Click **OK** or press **Enter**.

The VLOOKUP dialog box pops up (Figure 4.6). It prompts you for the Lookup_value, Table_array, Col_index_number, and the Range_lookup.

	A	B	C	D	E	F	G	H	I
2	Description	Quantity	Unit	$/Unit	Total $				
3	Shingles	28	SQ	B,3,False)	$ -				
4	Felt	7							
5	Drip Edge	18							
6	Ice & Water Shield	2							
7	Starter Strip	1							
8	Hip & Ridge	1.67							
9	Roofing Nails	56							
10	Total								
11									
12	Roofing Database								
13	Description		Unit						
14	25-Yr Asphalt Shingles		SQ	$					
15	30-Yr Asphalt Shingles		SQ	$					
16	30-Yr Architectural Shingles		SQ	$					
17	15# Felt		Roll	$ 10.95					
18	30# Felt		Roll	$ 11.95					
19	Drip Edge		EA	$ 3.85					
20	Ice & Water Shield		Roll	$ 68.50					
21	Starter Strip		SQ	$ 29.50					
22	Hip & Ridge		SQ	$ 29.50					
23	Plastic Caps		LB	$ 1.28					
24	Roofing Nails		LB	$ 0.78					

VLOOKUP dialog box shows: =VLOOKUP(A3,RoofingDB,3,False)
- Lookup_value: A3 = "Shingles"
- Table_array: RoofingDB = {"25-Yr Asphalt Shin...
- Col_index_num: 3 = 3
- Range_lookup: False = FALSE

Searches for a value in the leftmost column of a table, and then returns a value in the same row from a column you specify. By default, the table must be sorted in an ascending order.
Lookup_value is the value to be found in the first column of the table, and can be a value, a reference, or a text string.

Figure 4.6 The VLOOKUP Dalog Box

DEFINITION: Lookup_value is the value that VLOOKUP will search for in Column 1 of a specified database. In other words, the Lookup_value is the depends-on value; the value that VLOOKUP returns depends on the Lookup_value (A3 in this case). VLOOKUP captures whatever is entered into cell A3 and matches it to an item in the list of descriptions in the first column of the roofing database. If no match is found, then VLOOKUP returns the #N/A error value.

Click on the box next to **Lookup_value** in the dialog box and enter A3. You can type A3 or simply click on cell **A3** and A3 will automatically be entered into the **Lookup_value** box.

DEFINITION: Table_array tells VLOOKUP which database or table array to use. When looking up the cost per unit ($/Unit) for roofing, we want VLOOKUP to look in the roofing database. In this case, we named it RoofingDB.

In the box next to **Table_array**, type RoofingDB, or click **Insert/Name/Paste Name**, and select **RoofingDB**. Click **OK** or press **Enter**.

> **TIP**
>
> **Paste Name** comes in handy when you can't remember the name that you gave the database. Alternatively, if the name is long, it may save some time typing it in.

> **TIP**
>
> To save time, double click the left mouse button on a selected name or function. Double clicking performs two steps at once and is equal to selecting the item and pressing the Enter key.

DEFINITION: Col_index_number is the column number in the database where the information that you want VLOOKUP to find is stored. We want VLOOKUP to find the value of the cost per unit ($/Unit) for a specific item in the roofing database (RoofingDB). There are three columns in RoofingDB. The first column stores item descriptions, the second column holds the units and the third column stores the costs per unit ($/Unit).

Enter 3 in the box next to **Col_index_number**.

DEFINTION: Range_lookup tells VLOOKUP whether to find an exact match in column 1 of the database with the Lookup_value (the value in cell A3) or whether to pick the nearest value in column 1 of the database that is greater than theLookup_Value. If False is entered, VLOOKUP makes an exact match in column 1 of the database with the Lookup_value. If True is entered, VLOOKUP will select from column 1 in the database the next value greater than the Lookup_value. In estimating, you will always want an exact match so that your cost per unit is accurate.

Type False in the box next to **Range_lookup**.

Notice in the Edit box for cell E3 the formula

=VLOOKUP(A3,RoofingDB,3,FALSE)

that was created by the Function Wizard. You can type formulas yourself or you can use the Function Wizard to help. When you click in each box of the Function Wizard, you are prompted (near the bottom of the dialog box) for the correct input. At any time, if you don't understand how to create a formula, click on the **?** box in the bottom-left corner of the Function Wizard dialog box for help.

When you have completed entering the information into the **Function Wizard**, click **OK**. You will see the #N/A error because shingles does not match with anything in the first column of the roofing database.

Change shingles in cell **A3** to exactly match one of the descriptions in the first column of **RoofingDB**. If, for example, you enter 30-Yr Asphalt Shingles, **VLOOKUP** will return the value $34.40 into cell **D3** (Figure 4.7).

	A	B	C	D	E
	A3		=	30-Yr Asphalt Shingles	
2	Description	Quantity	Unit	$/Unit	Total $
3	30-Yr Asphalt Shingles	28	SQ	$ 34.40	$ 963.20
4	Felt	7	Roll		$ -
5	Drip Edge	18	EA		$ -
6	Ice & Water Shield	2	Roll		$ -
7	Starter Strip	1	SQ		$ -
8	Hip & Ridge	1.67	SQ		$ -
9	Roofing Nails	56	LB		$ -
10	Total				$ 963.20
11					
12	**Roofing Database**				
13	Description	Unit	$/Unit		
14	25-Yr Asphalt Shingles	SQ	$ 29.50		
15	30-Yr Asphalt Shingles	SQ	$ 34.40		
16	30-Yr Architectural Shingles	SQ	$ 39.85		
17	15# Felt	Roll	$ 10.95		
18	30# Felt	Roll	$ 11.95		
19	Drip Edge	EA	$ 3.85		
20	Ice & Water Shield	Roll	$ 68.50		
21	Starter Strip	SQ	$ 29.50		
22	Hip & Ridge	SQ	$ 29.50		
23	Plastic Caps	LB	$ 1.28		
24	Roofing Nails	LB	$ 0.78		

Figure 4.7 Correcting the Lookup_Value (Cell A3) to Match an Item in Column 1 of the Roofing Database

After you have tested the formula and it works the way that you want, copy the **VLOOKUP** formula in **D3** to cells **D4:D9**.

Again, you will notice an error value (#N/A) in cell D4 because felt does not match exactly with 15# Felt or with 30# Felt. Make the correction to cell **A4**. Type 15# Felt in **D4**.

You can also use **VLOOKUP** in Column C to look up the item unit (Unit). For example, in cell **A3** change **30-yr Ashalt Shingles** to **Ice & Water Shield**. The unit for **Ice & Water Shield** is **Roll**, not **SQ**. If the **VLOOKUP** formula were entered into **C3**, the value **Roll** would automatically be returned.

1. Select **C3**.
2. Enter the formula
=VLOOKUP(A3,RoofingDB,2,FALSE)

You can type the VLOOKUP formula yourself or you can use the Function Wizard to help. The only difference between the formula in C3 and the formula in D3 is that the Col_index_number is 2 instead of 3. The units are listed in column 2 in the roofing database. Test the formula and copy it to cells C4:C9.

Data Validation

The most time-consuming part of estimating is inputting data into the system. If you struggle with typing, you can use all the help you can get from the computer.

Fortunately, **Data Validation** can change data entry from the tedious chore of typing to one of clicking on an item in a list.

Earlier, in cell A3, you typed 30-Yr Asphalt Shingles and in cell A4, 15# Felt. Say you changed your mind and instead wanted 30-Yr Architectural Shingles and 30# Felt. You would need to retype the descriptions into cells A3 and into A4. We have already begun the process of setting up to use Data Validation with the roofing descriptions. Remember when we were setting up the roofing database, we named the cell range A14:C24 RoofingDB? We then named range A14:A24 RoofingList. We will use RoofingList with Data Validation.

1. Select **A3**.
2. Click **Data** on the **Menu** bar.
3. Click **Validation**.

There are several validation criteria (Figure 4.8). Validation is a method to control what information goes into a cell. In our case we want to specify information that is on a list. Select **List**.

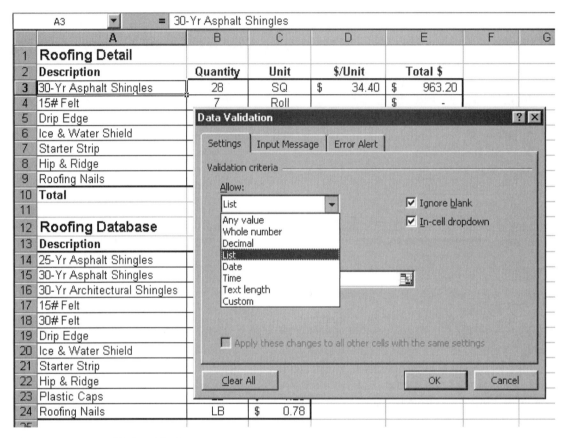

Figure 4.8 Data Validation Criteria

The Data Validation dialog box asks for the source of the list (Figure 4.9). In the box under source, enter the range of cells where the list is located. You already named the range of cells (RoofingList) where the list of roofing descriptions is located.

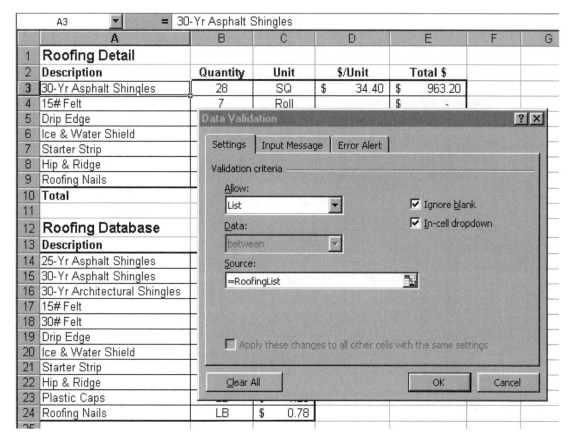

Figure 4.9 Entering a Data Validation List Source

In the box under **Source**, type =RoofingList. If you do not remember the name of the list, click **Insert/Name/Paste Name** and select **RoofingList**.

An alternative method to specify the source for the list is to click in the **Source** box and then select the cells that contain the list. This method only works if the list is located on the same worksheet as the database. If the list is not located on the same worksheet, the list has to be given a name and the name is used in the **List Source** box.

When you have finished entering the validation criteria, click **OK**. You will now notice a small down-facing arrow next to cell **A3**.

Click on the arrow and a drop-down list of roofing descriptions will appear (Figure 4.10).

	A	B	C	D	E
1	**Roofing Detail**				
2	Description	Quantity	Unit	$/Unit	Total $
3	30-Yr Asphalt Shingles	28	SQ	$ 34.40	$ 963.20
4	25-Yr Asphalt Shingles	7	Roll	$ 10.95	$ 76.65
5	30-Yr Asphalt Shingles	18	EA	$ 3.85	$ 69.30
6	30-Yr Architectural Shingles	2	Roll	$ 68.50	$ 137.00
7	15# Felt	1	SQ	$ 29.50	$ 29.50
8	30# Felt	1.67	SQ	$ 29.50	$ 49.27
9	Drip Edge	56	LB	$ 0.78	$ 43.68
	Ice & Water Shield				
	Starter Strip				
10	Total				$ 1,368.60
11					
12	**Roofing Database**				
13	Description	Unit	$/Unit		
14	25-Yr Asphalt Shingles	SQ	$ 29.50		
15	30-Yr Asphalt Shingles	SQ	$ 34.40		
16	30-Yr Architectural Shingles	SQ	$ 39.85		
17	15# Felt	Roll	$ 10.95		
18	30# Felt	Roll	$ 11.95		
19	Drip Edge	EA	$ 3.85		
20	Ice & Water Shield	Roll	$ 68.50		
21	Starter Strip	SQ	$ 29.50		
22	Hip & Ridge	SQ	$ 29.50		
23	Plastic Caps	LB	$ 1.28		
24	Roofing Nails	LB	$ 0.78		

Figure 4.10 Data Validation Drop-Down List

Select **30-Yr Architectural Shingles** and this shingle description will automatically be entered into cell **A3** (Figure 4.11). Notice the changes that are made automatically in the Unit and $/Unit columns.

	A	B	C	D	E
1	**Roofing Detail**				
2	Description	Quantity	Unit	$/Unit	Total $
3	30-Yr Architectural Shingles	28	SQ	$ 39.85	$ 1,115.80
4	15# Felt	7	Roll	$ 10.95	$ 76.65
5	Drip Edge	18	EA	$ 3.85	$ 69.30
6	Ice & Water Shield	2	Roll	$ 68.50	$ 137.00
7	Starter Strip	1	SQ	$ 29.50	$ 29.50
8	Hip & Ridge	1.67	SQ	$ 29.50	$ 49.27
9	Roofing Nails	56	LB	$ 0.78	$ 43.68
10	Total				$ 1,521.20
11					
12	**Roofing Database**				
13	Description	Unit	$/Unit		
14	25-Yr Asphalt Shingles	SQ	$ 29.50		
15	30-Yr Asphalt Shingles	SQ	$ 34.40		
16	30-Yr Architectural Shingles	SQ	$ 39.85		
17	15# Felt	Roll	$ 10.95		
18	30# Felt	Roll	$ 11.95		

Figure 4.11 Using Data Validation Drop-Down Lists to Select Items

Copy cell **A3** to cells **A4** through **A9**. Choose new item descriptions from the drop-down list for cells **A4:A9**.

The corresponding units and unit prices will automatically be updated. The prices will be extended and totaled instantly.

VLOOKUP can be an especially useful function for items such as concrete, a material that is used in several detail sheets. When you receive price increases from the concrete supplier, you need only update the concrete prices in one database and all of the concrete will reflect the new concrete prices.

Now, when you do a roofing estimate, all you do is select the items you want to take off and enter the quantities. The units and the costs per unit will be automatically and instantly updated. This is especially useful when making change orders. No longer do you need to guess at what to charge a customer for a requested change.

Summary

You can create a roofing detail sheet and a roofing database of your own. If you have difficulty, you may use the sample Excel file **(Chapter 4.xls)** that is provided on the CD that accompanies this book. **Chapter 4.xls** starts at the beginning of this chapter and **Chapter 4B.xls** has all of the completed formulas that were used in this chapter.

Linking

CHAPTER 5

IN THIS CHAPTER

- Use the IF function to create IF, Then, Else decision formulas.
- Switch between viewing values and formulas on the worksheet.
- Link spreadsheets with other worksheets, with other workbooks, and with other files.
- Hyper link to move quickly to another cell in the same worksheet.

Chapter 3 focused on developing a Cost Breakdown Summary sheet capable of summarizing all of the costs of a project. Chapter 4 focused on creating a roofing detail sheet and a roofing database. The detail sheet contains the quantity take-off detail of the items that are included in the roofing estimate. It also contains formulas that automate most of the estimating process. The database that you created supports the detail sheet by making data available through VLOOKUP formulas. When suppliers or trade contractors change the prices that they charge, you can make those changes in the database and the changes are automatically updated in the detail sheet.

Have you used spreadsheets or other software programs that, when you finished with a specific task, you had to copy all of the information and paste it to the new location or to another program? Worse, still, is having to re-type information that can't be copied through the computer. Having to copy and paste, or retype information can be frustrating and very time-consuming. And one thing that's true about most builders I know is that they don't have a lot of extra time on their hands.

In this chapter you will learn methods that will allow your spreadsheets to interact with other worksheets, with other workbooks, and with other files. The Excel file on the CD that accompanies this book, **Chapter 5 Linking.xls**, has three worksheets; the first is the Cost Breakdown Summary sheet that was created in chapter three, the second contains the roofing detail, and the third is the roofing database. This layout (with the database on a separate sheet) more accurately models a finished estimating spreadsheet. Another option would be to place the database in different workbook.

The IF Function

Before we deal with linking cells, there are a few things we need to do to fix the Roofing Detail sheet. First, select the Roofing Detail sheet in the **Chapter 5 Linking.xls** file. You will notice that several extra rows have been added to allow for additional roofing items. Also notice the #N/A error in many of the cells. Remember the VLOOKUP functions that were entered in Columns C and D? These VLOOKUP functions are looking to Column A for the Lookup_value, however, because the **Lookup_value** in cells A10:A16 are blank, VLOOKUP cannot make a match in the roofing database and therefore, returns the error values #N/A. Not only do the #N/A error signs detract from the appearance of the worksheet, but they don't allow the Roofing Total to be calculated (Figure 5.1).

	A	B	C	D	E
1	Roofing Detail				
2	Description	Quantity	Unit	$/Unit	Total $
3	30-Yr Architectural Shingles	28	SQ	$ 39.85	$ 1,115.80
4	30# Felt	7	Roll	$ 11.95	$ 83.65
5	Drip Edge	18	EA	$ 3.85	$ 69.30
6	Ice & Water Shield	2	Roll	$ 68.50	$ 137.00
7	Starter Strip	1	SQ	$ 29.50	$ 29.50
8	Hip & Ridge	1.67	SQ	$ 29.50	$ 49.27
9	Roofing Nails	56	LB	$ 0.78	$ 43.68
10			#N/A	#N/A	#N/A
11			#N/A	#N/A	#N/A
12			#N/A	#N/A	#N/A
13			#N/A	#N/A	#N/A
14			#N/A	#N/A	#N/A
15			#N/A	#N/A	#N/A
16			#N/A	#N/A	#N/A
17	Total				#N/A
18					

Figure 5.1 Without a Description in Column A, VLOOKUP Returns the #N/A Error Sign

We could state, in words, the solution to our problem this way, if the item description in A3 is blank, then C3 should be blank, otherwise, enter the result from the **VLOOKUP** formula in **C3**. Whenever you state the solution to a problem in words and the sentence starts with "if", it is likely that you can use the IF function in Excel to solve it.

The **IF** function is a conditional statement that takes the form

=IF(Logical Test, Value if True, Value if False)

Which means if the logical test is true, then do something; otherwise (if false) do something else. To create a formula using the IF function:

Enter =IF(A3="","",VLOOKUP(A3,RoofingDB,2,False)) into cell **C3**.
Copy the formula to cells **C4:C16**.

This same formula can also be used in the cost per unit column (column D).

Enter =IF(A3="","",VLOOKUP(A3,RoofingDB,3,False)) into cell **D3**.

Notice that the Col_index_number , 2, has been replaced with 3, because the cost per unit ($/Unit) is found in column 3 of RoofingDB.

Copy the formula to cells **D4:D16**.

Notice that the VLOOKUP function is nested inside the Value if False parameter of the IF Statement. Nesting formulas inside of one another can be very powerful. Nesting formulas not only enables you do accomplish several functions in one, but the formulas are made more powerful because of the combinations of functions.

The #N/A error in the item Total Cost column (column E) can also be corrected using the IF function. One formula that could be used to solve the problem is shown:

Enter =IF(B3= "",IF(D3="","",B3*D3) into cell **E3**.

The interpretation of this formula is: if B3 is blank, or if D3 is blank, then E3 will be blank. If neither B3 nor D3 are blank the product of the formula B3 *D3 (quantity times unit price) will be returned to cell E3. In this formula, one IF statement is nested within the other IF statement. As many as seven IF statements can be nested within each other.

Copy the formula to cells **E4:E16**.

Viewing the Formulas in Cells

You can switch between viewing values and formulas on the worksheet.

While holding down the **Ctrl**, press the grave accent (`` ` ``), which is located above the tab key. **Ctrl+`** is a toggle key that will switch between the formula view and the value view.

If you entered the formulas correctly, your worksheet formulas should look like Figure 5.2.

	C	D	E
2	Unit	$/Unit	Total $
3	=IF(A3="","",VLOOKUP(A3,RoofingDB,2,FALSE))	=IF(A3="","",VLOOKUP(A3,RoofingDB,3,FALSE))	=IF(B3="","",IF(D3="","",B3*D3))
4	=IF(A4="","",VLOOKUP(A4,RoofingDB,2,FALSE))	=IF(A4="","",VLOOKUP(A4,RoofingDB,3,FALSE))	=IF(B4="","",IF(D4="","",B4*D4))
5	=IF(A5="","",VLOOKUP(A5,RoofingDB,2,FALSE))	=IF(A5="","",VLOOKUP(A5,RoofingDB,3,FALSE))	=IF(B5="","",IF(D5="","",B5*D5))
6	=IF(A6="","",VLOOKUP(A6,RoofingDB,2,FALSE))	=IF(A6="","",VLOOKUP(A6,RoofingDB,3,FALSE))	=IF(B6="","",IF(D6="","",B6*D6))
7	=IF(A7="","",VLOOKUP(A7,RoofingDB,2,FALSE))	=IF(A7="","",VLOOKUP(A7,RoofingDB,3,FALSE))	=IF(B7="","",IF(D7="","",B7*D7))
8	=IF(A8="","",VLOOKUP(A8,RoofingDB,2,FALSE))	=IF(A8="","",VLOOKUP(A8,RoofingDB,3,FALSE))	=IF(B8="","",IF(D8="","",B8*D8))
9	=IF(A9="","",VLOOKUP(A9,RoofingDB,2,FALSE))	=IF(A9="","",VLOOKUP(A9,RoofingDB,3,FALSE))	=IF(B9="","",IF(D9="","",B9*D9))
10	=IF(A10="","",VLOOKUP(A10,RoofingDB,2,FALSE))	=IF(A10="","",VLOOKUP(A10,RoofingDB,3,FALSE))	=IF(B10="","",IF(D10="","",B10*D10))
11	=IF(A11="","",VLOOKUP(A11,RoofingDB,2,FALSE))	=IF(A11="","",VLOOKUP(A11,RoofingDB,3,FALSE))	=IF(B11="","",IF(D11="","",B11*D11))
12	=IF(A12="","",VLOOKUP(A12,RoofingDB,2,FALSE))	=IF(A12="","",VLOOKUP(A12,RoofingDB,3,FALSE))	=IF(B12="","",IF(D12="","",B12*D12))
13	=IF(A13="","",VLOOKUP(A13,RoofingDB,2,FALSE))	=IF(A13="","",VLOOKUP(A13,RoofingDB,3,FALSE))	=IF(B13="","",IF(D13="","",B13*D13))
14	=IF(A14="","",VLOOKUP(A14,RoofingDB,2,FALSE))	=IF(A14="","",VLOOKUP(A14,RoofingDB,3,FALSE))	=IF(B14="","",IF(D14="","",B14*D14))
15	=IF(A15="","",VLOOKUP(A15,RoofingDB,2,FALSE))	=IF(A15="","",VLOOKUP(A15,RoofingDB,3,FALSE))	=IF(B15="","",IF(D15="","",B15*D15))
16	=IF(A16="","",VLOOKUP(A16,RoofingDB,2,FALSE))	=IF(A16="","",VLOOKUP(A16,RoofingDB,3,FALSE))	=IF(B16="","",IF(D16="","",B16*D16))
17			=SUM(E3:E16)

Figure 5.2 Viewing the Formulas in a Worksheet

Press **Ctrl+`** again and you will be able to view the values again on the Roofing Detail sheet (Figure 5.3).

	A	B	C	D	E
1	**Roofing Detail**				
2	Description	Quantity	Unit	$/Unit	Total $
3	30-Yr Architectural Shingles	28	SQ	$ 39.85	$ 1,115.80
4	30# Felt	7	Roll	$ 11.95	$ 83.65
5	Drip Edge	18	EA	$ 3.85	$ 69.30
6	Ice & Water Shield	2	Roll	$ 68.50	$ 137.00
7	Starter Strip	1	SQ	$ 29.50	$ 29.50
8	Hip & Ridge	1.67	SQ	$ 29.50	$ 49.27
9	Roofing Nails	56	LB	$ 0.78	$ 43.68
10	Plastic Caps		LB	$ 1.28	
11	15# Felt				
12	30# Felt				
13	Drip Edge				
	Ice & Water Shield				
14	Starter Strip				
15	Hip & Ridge				
	Plastic Caps				
16	Roofing Nails				
17	**Total**				$ 1,528.20

Figure 5.3 The Roofing Detail Sheet–Values View

Figure 5.3 also shows the Data Validation drop-down box that can be used to quickly select take-off items on the detail sheet. When an item is selected, the unit and unit price (cells C10 and D10) are automatically entered. When the estimator enters the quantity, the item total is automatically calculated and the roofing total (E17) is instantly updated.

Linking

It would save time if the Roofing cost on the summary sheet (in the Estimated-Cost column next to Roofing) could automatically be updated to match the roofing total ($1,528.20) on the Detail sheet when it changed. When a roofing quantity or price was changed, the new roofing total would instantly appear next to Roofing on the Cost Breakdown Summary sheet. Linking is an easy and powerful way to replicate the value of one cell to another cell. Linking causes the value in the linked (destination) cell to duplicate the value in the original (source) cell.

There are two ways to create a link.

Method One

The easiest way is to select the target (destination) cell.

1. Select cell **E40** of the Cost Breakdown Summary sheet (Figure 5.4).
2. Enter the = sign.
3. Click on the **RoofingDetail** sheet tab to activate the Roofing Detail sheet.
4. Click on the cell that contains the value for the roofing total (**E17**, the source). When you click on the cell, Excel automatically inserts the cell address into the formula.
5. Press **Enter** to complete the formula.
6. Test your link by changing a quantity or a cost per unit ($/unit) on the Roofing Detail sheet and checking the summary sheet to verify the update.

	E40		=	=RoofingDetail!E17	
	D		E		F
36	Heating				
37	Air Conditioning				
38	Electrical				
39	Light Fixture Allow.				
40	Roofing		$	1,528.20	
41	Insulation				
42	Drywall				

Figure 5.4 Creating a Link

Method Two

There is another way to create links:

1. Copy the source information (on the Roofing Detail sheet, copy cell **E17**).
2. Move to the Cost Breakdown Summary sheet and select cell **E40**.
3. Click **Edit**.
4. Click **Paste Special**.
5. Click **Paste Link**.

> **TIP**
>
> A quick way to toggle the $ signs on and off for a reference is to press the **F4** function key. In the **Edit** bar, place the cursor in the reference and press **F4**. Continue to press **F4** and watch the changes to the formula.

> **NOTE**
>
> Notice the $ signs in the formula where Paste Link was used (Figure 5.5). The $ sign signifies an absolute reference. If cell E40 is copied down one cell, the formula would still refer to RoofingDetail!$E$17. Without the $ signs, the reference in the formula is relative, that is if the formula were copied down one cell, the formula in the new cell would refer to RoofingDetail!$E$18.

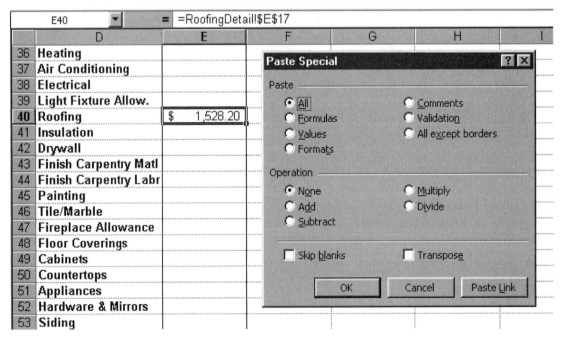

Figure 5.5 Paste Link Can Be Used to Create a Link

Hyperlink

It has been simple, so far, to move from sheet to sheet by clicking the sheet tabs at the bottom of the worksheets. As you add new sheets to your estimating program, it becomes more cumbersome and slow to move back and forth between sheets. Using hyperlink can solve this problem.

Clicking hyperlinked text or graphics causes the focus of Excel to jump to a new location. When you click on the hyperlinked text or graphic, the referenced cell is automatically activated.

Hyperlink Example

1. Select cell **D40** on the Cost Breakdown Summary sheet (Figure 5.6).

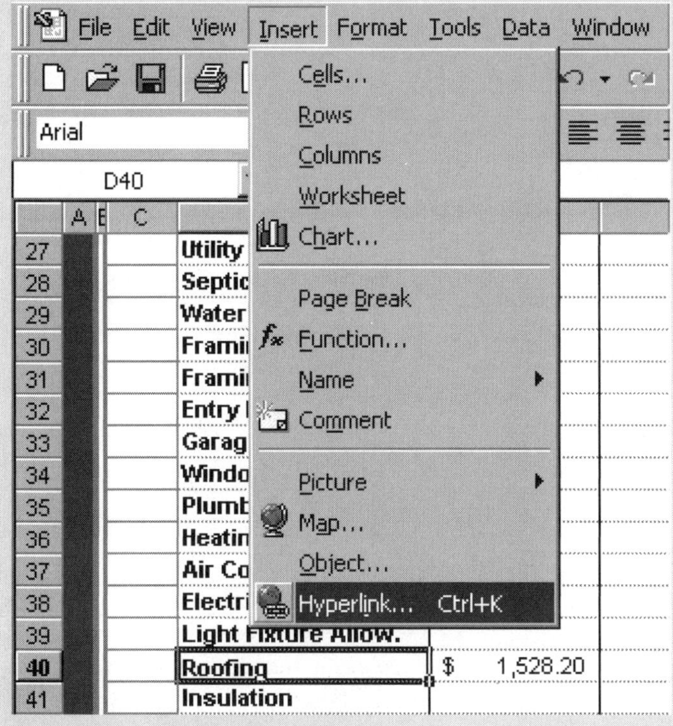

Figure 5.6 Inserting Hyperlinks

Continued

Hyperlink Example *Continued*

2. Click **Insert/Hyperlink**. The **Edit Hyperlink** dialog box will appear (Figure 5.7).

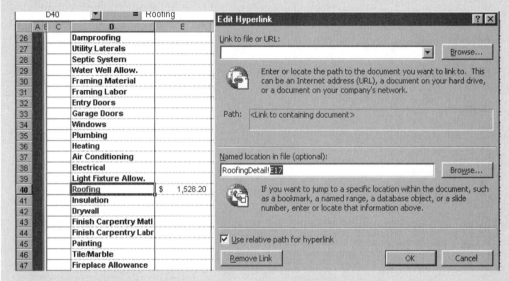

Figure 5.7 Edit Hyperlink Dialog Box

3. Click in the **Named location in file** box.
4. Enter the destination location (the location to where you want Excel to jump).
5. Type the name of the sheet, followed by the exclamation mark (**!**) and the cell address (for example, type RoofingDetail!E17).

A quick way to enter the name of the destination reference location is to:

1. Click **Browse** for a pop-up list of worksheets in the workbook (Figure 5.8).
2. Pick **RoofingDetail**
3. In the **Reference** box enter the cell location, **E17**.
4. Click **OK** and the link will be created.
5. Click on the newly created Roofing link to see it work.

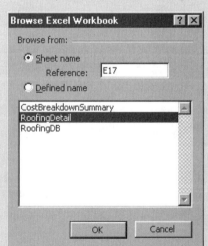

Figure 5.8 The Browse Excel Workbook Dialog Box

You can also create hyperlinks to named cells and named ranges, for example, suppose the total amount on the Roofing detail sheet (**E17**, Figure 5.9) was named RoofingTotal.

RoofingTotal		=SUM(E3:E16)		
A	**B**	**C**	**D**	**E**
1 Roofing Detail				
2 Description	**Quantity**	**Unit**	**$/Unit**	**Total $**
3 30-Yr Architectural Shingles	28	SQ	$ 39.85	$ 1,115.80
4 30# Felt	7	Roll	$ 11.95	$ 83.65
5 Drip Edge	18	EA	$ 3.85	$ 69.30
6 Ice & Water Shield	2	Roll	$ 68.50	$ 137.00
7 Starter Strip	1	SQ	$ 29.50	$ 29.50
8 Hip & Ridge	1.67	SQ	$ 29.50	$ 49.27
9 Roofing Nails	56	LB	$ 0.78	$ 43.68
10 Plastic Caps		LB	$ 1.28	
11				
12				
13				
14				
15				
16				
17 Total				$ 1,528.20
18				

Figure 5.9 Cell Address E17 Named RoofingTotal

To create a hyperlink reference to a named cell:

1. Select the cell (**D40**, Figure 5.7) that will hyperlink (launch) to a new location.
2. Click **Insert**.
3. Click **Hyperlink**.
4. Click the **Browse** button next to the **Named location in file (optional)**: box. The Browse Excel Workbook dialog box pops up.
5. Click the **Defined** name option button (Figure 5.10).
6. Double-click on **RoofingTotal** and the name will automatically be entered into the **Named location in file (optional)**: box.

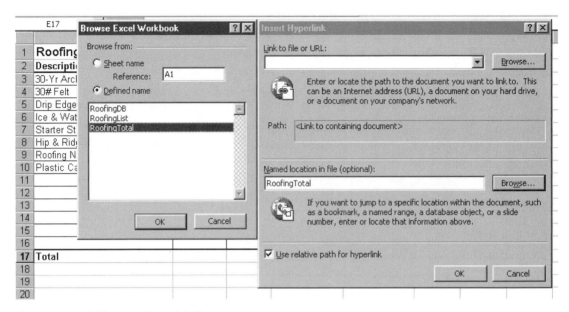

Figure 5.10 Linking to a Named Cell

The advantage to hyperlinking using a named cell (RoofingTotal) as a reference rather than using an address reference (RoofingDetail!E17), is that when the cell is moved, such as when rows above are inserted or deleted above the cell location, the hyperlink reference will adjust to the new position of the named cell.

You can also create hyperlinks to other files. Suppose you had a document file with information on the specifications of your house plans, and you wanted to be able to access the file from inside Excel. To open the specifications file from inside Excel:

1. Click in the **Link to file or URL** box (Figure 5.11 upper right).
2. Enter the path (location) and name of the file that you want to open. If the file is named Specifications.doc, its path is C:\My Documents\Plans\. Type C:\My Documents\Plans\Specifications.doc.

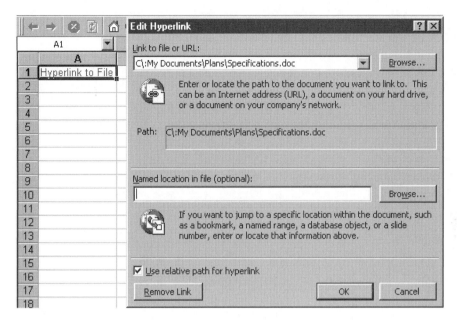

Figure 5.11 Hyperlink to File Specifications.doc

Links can also be created to the Internet. If you needed information about local chapters of NAHB or other information:

1. Click in the **Link** to file or **URL** box (Figure 5.10 upper right).
2. Enter the destination location address, http:\\www.nahb.org.

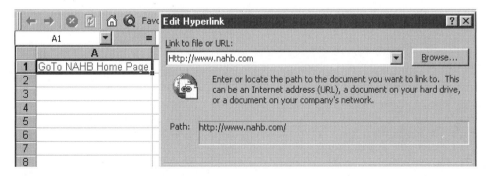

Figure 5.12 Hyperlink to NAHB Website

You can also type the file path and name or use Browse to select the file. The path to the file and the file name will automatically be entered into the Link to file or URL box.

This chapter discussed how to use the IF function to create IF, Then, Else decision formulas. In this chapter you learned how to link cells. The chapter also focused on hyper linking. Hyper linking can be used to move quickly to another cell in the same worksheet, to a cell in a different worksheet, or to another file, or Internet source.

If you have correctly completed the instructions in this chapter, you will have created a file similar to **Chapter 5B Linking.xls**, which is located on the CD that accompanies this book. Remember to use the **Help/Contents** and **Index** feature if you get stuck on any specific concept.

Summary

Linking connects all of your work by bringing together all of the loose ends. Through linking, information is forwarded to the areas where additional processing can occur. Linking also makes summarizing information very fast and accurate. The next chapter will introduce you to formulas and functions that you can use to automate the calculation of quantities for items such as concrete, rebar, permit fees, roofing, etc. By adapting the formulas you learn in Chapter 6, you will be able to write formulas to calculate quantities for most work items in your estimating spreadsheets.

CHAPTER 6

More on Formulas and Functions

IN THIS CHAPTER

Set up detailed estimating worksheets for:

- Concrete
- Rebar
- Permit Fees
- Connection and Impact Fees
- Roofing

We have seen the remarkable power that functions such as **VLOOKUP**, **IF**, **SUM**, and **NOW** add to the formulas that drive and automate your spreadsheets. In this chapter, we will look at some new functions and formulas that will increase the functionality of your spreadsheets and improve the speed with which you can complete your estimates. You will learn specific applications to improve your spreadsheet estimating. You will also learn new ways to apply some of the functions from chapters four and five.

During the creation of these detailed estimating sheets, you will learn how to implement the following functions, formulas and methods, which can greatly improve the speed and accuracy of your estimates.

Cubic Yard Formula	to calculate concrete quantities.
Factoring Waste	adding waste factors to formulas
ROUNDUP function	to round concrete quantities to the nearest ¼ yard.
The IF function	to control input into cells.
More on VLOOKUP	to lookup information from the database.
MATCH function	to lookup costs from more than one vendor.
List Validation	for selecting vendors or subcontractors.
Rebar Formulas	to automatically calculate rebar quantities.
Custom formatting	for improved readability.
Nested IF statements	to make complex decisions.
Adding Comments	to give instructions to other users.
Slope Formulas	to calculate the roofing slope factors.
Roofing Formulas	to calculate quantities of shingles.
Estimating Methods	sub-bid versus detail estimating.

Estimating Concrete

Estimating concrete presents some unique challenges when setting up a spreadsheet. Although electronic spreadsheets can reduce much of the drudgery of estimating, using them still requires input from the estimator. The trick is to minimize the input while maintaining accuracy. When creating computer-estimating spreadsheets, design them so that the only inputs from the estimator are the take-off units.

When dealing with concrete flatwork, for example, many estimators input the cubic yards of concrete and let the computer calculate the total cost by multiplying the cubic yards by the cost per cubic yard. Quantities, such as cubic yards (CY), are not take-off quantities, but calculated quantities. If spreadsheets are designed so that only take-off quantities are entered (Figure 6.1), then estimators are not required the intermediate step of calculating the volume. An added benefit is that records of the take-off details are preserved for verifying amounts and for future references.

More on Formulas and Functions 73

	A	B	C	D	E	F	G	H	I
1	**Flatwork Detail**								
2	Description	Bag Mix	Thick "	Width '	Length '	SF	CY	$/Unit	Total $
3	Floor Slab	5.5	4	32	58				
4	Garage Floor	5.5	4	24	28				
5	Driveway	6	4	22	30				
6	Driveway Apron	7	6	4	22				
7									
8									
9	**Concrete DB**					Take-off Quantities			
10	Bag Mix	Unit	$/ Unit						
11	5	CY	$67.00						
12	5.5	CY	$69.00						
13	6	CY	$71.00						
14	6.5	CY	$73.00						
15	7	CY	$75.00						
16									

Figure 6.1 Restricting Spreadsheet Input to Take-Off Quantities

Enter the take-off quantities as shown in Figure 6.1. In this concrete flatwork take-off, SF and CY are both calculated quantities that the computer can instantly calculate. Square feet (SF) is needed for pricing labor and Cubic yards (CY) is required for pricing concrete. The cost per Unit ($/Unit) is a lookup value and the total cost (Total $) is a calculated value, both of which can be processed automatically by the computer.

The formula to calculate SF is width, in feet, multiplied by length, in feet. Enter the SF formula in cell F3:

1. Select cell **F3**.
2. Type =D3*E3.
3. Copy the formula to cells **F4:F6**. Either copy cell **F3**, select cells **F4:F6** and click the **Paste** button, or select the **Fill** handle on cell **F3** and drag it down to cell **F6**.

The formula for CY is thickness, in feet, times width, in feet, times length, in feet, divided by 27, since there are 27 cubic feet in a cubic yard (3 ft3 =3 x 3 x 3 = 27). Because our input for thickness is given in inches, the thickness needs to be converted to feet by dividing the inches by 12.

1. Select cell **G3**.
2. Type =C3/12*D3*E3/27.

3. Copy **G3** to cells **G4:G6**. The equivalent formula, =(C3/12*D3*E3)/27, will produce the same result but may be easier to read. You could also use the equivalent formula that would multiply the flatwork thickness in feet (divide the thickness in inches by 12) by the square footage, and divide by 27 (=C3/12*F3/27).

Factoring Waste

Notice the results of the CY calculations in cells G3:G6 (Figure 6.2). Two things are off the mark. First, there has been no accounting for waste, and second, concrete can typically be purchased in ¼-yard increments. To solve the waste problem, we could simply add a waste factor into the CY formula, =C3/12*D3*E3/27*1.05. It would improve the design, however, to allow the user to change the factor. It is not a good idea to treat variables, such as the waste factor, as littorals inside the formula. One reason that you wouldn't want to put the waste factor value inside the formula is that it is more difficult to check the formula to make sure it has the correct waste factor. Another reason is that if you want to change the waste factor, all of the CY formulas have to be changed.

G3		=	=C3/12*D3*E3/27				
	A	B	C	D	E	F	G	H
1	**Flatwork Detail**							
2	Description	Bag Mix	Thick "	Width '	Length '	SF	CY	$/Unit
3	Floor Slab	5.5	4	32	58	1856	22.914	
4	Garage Floor	5.5	4	24	28	672	8.2963	
5	Driveway	6	4	22	30	660	8.1481	
6	Driveway Apron	7	6	4	22	88	1.6296	
7								

Figure 6.2 Cubic Yard Formulas

Create a cell (G1, Figure 6.3) that prompts the user to input the waste factor.

G3		=	=C3/12*D3*E3/27*(1+FlatworkWasteFactor)				
	A	B	C	D	E	F	G	H
1	**Flatwork Detail**				Waste Factor:		5%	
2	Description	Bag Mix	Thick "	Width '	Length '	SF	CY	$/Unit
3	Floor Slab	5.5	4	32	58	1856	24.059	
4	Garage Floor	5.5	4	24	28	672	8.7111	
5	Driveway	6	4	22	30	660	8.5556	
6	Driveway Apron	7	6	4	22	88	1.7111	
7								

Figure 6.3 Adding a Waste Factor to the Cubic Yard Formula

1. Format **G1** for percentage (click the **%** button).
2. Type the waste factor.
3. Name the cell **FlatworkWasteFactor** (click on the cell and type FlatworkWasteFactor in the **Name** box.)
4. In the CY formula, add *(1+FlatWorkWasteFactor), the completed formula should look like =C3/12*D3*E3/27*(1+FlatWorkWasteFactor).

Test the formula by changing the waste factor in cell **G1** and verify the results.

Rounding Numbers

The waste factor was added to the item total costs but the quantities are not yet rounded to one-quarter cubic yard increments. The **ROUNDUP** function can help us with this problem. ROUNDUP will round the results up to the nearest whole number or any number of digits that you choose. The syntax or form of the ROUNDUP function is:

ROUNDUP(number,num_digits)

The **num_digits** is optional. If left off or set at zero, the number is rounded up to the nearest integer.

Example: =ROUNDUP(27.374,) returns the value 28
=ROUNDUP(27.374,0) returns the value 28

If **num_digits** is greater than 0, then the number is rounded up to the specified number of decimal places.

Example: =ROUNDUP(27.374,2) returns the value 27.38

If **num_digits** is less than 0, then number is rounded up to the left of the decimal.

Example: =ROUNDUP(27.374,-1) returns the value 30

To round up cubic yards of concrete to the nearest one-quarter CY takes a little trickery. You have to multiply the number by 4, then round up to the nearest integer, then divide the result by 4.

Example: =ROUNDUP(27.374*4,0)/4 returns the value 27.5

In our flatwork example (Figure 6.4), the CY amount (number) is represented by C3/12*D3*E3/27*(1+FlatworkWasteFactor) . If we plug this number into the ROUNDUP formula (=ROUNDUP(number*4,0)/4,) the resulting formula is:

=ROUNDUP(C3/12*D3*E3/27*(1+FlatworkWasteFactor)*4,0)/4

The outcome of copying this formula to cells **G3:G6** can be seen in Figure 6.4.

G3			=	=ROUNDUP(C3/12*D3*E3/27*(1+FlatworkWasteFactor)*4,0)/4			
A	B	C	D	E	F	G	H
1 **Flatwork Detail**					**Waste Factor:**	5%	
2 Description	Bag Mix	Thick "	Width '	Length '	SF	CY	$/Unit
3 Floor Slab	5.5	4	32	58	1856	24.25	
4 Garage Floor	5.5	4	24	28	672	8.75	
5 Driveway	6	4	22	30	660	8.75	
6 Driveway Apron	7	6	4	22	88	1.75	
7							

Figure 6.4 Using the ROUNDUP Function to Round Up to Quarter-Yard Increments

VLOOKUP Revisited

The cost per unit ($/Unit) is a lookup value that depends on the bag mix (sack mix) of the concrete.

1. In cell **H3** enter the formula=VLOOKUP(B3,ConcreteDB,3,False) (see Chapter 4 for a review of the **VLOOKUP** function).
2. Copy the formula to cells **H4:H6** by dragging **H3's** fill handle to cell **H6**.
3. The **VLOOKUP** formula in cell **H3** (Figure 6.5) captures the **Lookup_Value** (5.5 in cell **B3**) and locates the matching reference in the table array, **ConcreteDB**. The value in the third column ($69.00) is returned to the $/Unit position in **H3**.

	A	B	C	D	E	F	G	H
					=VLOOKUP(B3,ConcreteDB,3,FALSE)			
1	Flatwork Detail					Waste Factor:	5%	
2	Description	Bag Mix	Thick "	Width '	Length '	SF	CY	$/Unit
3	Floor Slab	5.5	4	32	58	1856	24.25	$ 69.00
4	Garage Floor	5.5	4	24	28	672	8.75	$ 69.00
5	Driveway	6	4	22	30	660	8.75	$ 71.00
6	Driveway Apron	7	6	4	22	88	1.75	$ 75.00
7								
8				Lookup_Value				
9	Concrete DB							
10	Bag Mix		Unit	$/ Unit				
11	5		CY	$67.00				
12	5.5		CY	$69.00				
13	6		CY	$71.00				
14	6.5		CY	$73.00				
15	7		CY	$75.00				
16	Column 1		Col 2	Col 3				
17								
18				ConcreteDB (Cells A11:C15)				

Figure 6.5 VLOOKUP Function

To calculate the item totals, type =G3*H3 into cell **I3**. Copy the formula in **I3** to **I4:I6**.

If B3 is left blank, however, then the #N/A error will display in I3 (Figure 6.6). With an error in I3, the total sum (I7) also displays an error value. If B3 is blank, we want the extended cost (I3) to be blank, or else we want the product of G5*H5. This problem can be solved by using the IF function as part of the total cost formula.

	A	B	C	D	E	F	G	H	I
			=G3*H3						
1	Flatwork Detail					Waste Factor:	5%		
2	Description	Bag Mix	Thick "	Width '	Length '	SF	CY	$/Unit	Total $
3	Floor Slab		4	32	58	1856	24.25	#N/A	#N/A
4	Garage Floor	5.5	4	24	28	672	8.75	$ 69.00	$ 603.75
5	Driveway	6	4	22	30	660	8.75	$ 71.00	$ 621.25
6	Driveway Apron	7	6	4	22	88	1.75	$ 75.00	$ 131.25
7									#N/A

Figure 6.6 Extending the Total Costs

1. Select **cell I3** (Figure 6.7).
2. Type =IF(B3="","",G3*H3).
3. Copy the formula to cells **I4:I6**.

	A	B	C	D	E	F	G	H	I
1	Flatwork Detail					Waste Factor:	5%		
2	Description	Bag Mix	Thick "	Width '	Length '	SF	CY	$/Unit	Total $
3	Floor Slab		4	32	58	1856	24.25	#N/A	
4	Garage Floor	5.5	4	24	28	672	8.75	$ 69.00	$ 603.75
5	Driveway	6	4	22	30	660	8.75	$ 71.00	$ 621.25
6	Driveway Apron	7	6	4	22	88	1.75	$ 75.00	$ 131.25
7									$1,356.25
8									

Cell I3 formula: =IF(B3="","",G3*H3)

Figure 6.7 Controlling the Display of Error Signs

You can see that while B3 is blank, the #N/A error displays in H3 to alert the estimator to enter the bag mix in cell B3. The #N/A error does not display in cell I3 and therefore, does not effect the total of the other cells. The formula for the total sum (I7) is =SUM(I3:I6).

Using the MATCH Function

In your business, you likely deal with more than one concrete supplier. Each concrete supplier has different unit prices that they charge. In the Concrete Database (ConcreteDB, Figure 6.8) there are three concrete suppliers to choose from: Rocktec, Redimix, and Hardcrete.

	A	B	C	D	E	F	G	H	I
1	Flatwork Detail					Waste Factor:	5%		
2	Description	Bag Mix	Thick "	Width '	Length '	SF	CY	$/Unit	Total $
3	Floor Slab	5.5	4	32	58	1856	24.25	$ 69.00	$1,673.25
4	Garage Floor	5.5	4	24	28	672	8.75	$ 69.00	$ 603.75
5	Driveway	6	4	22	30	660	8.75	$ 71.00	$ 621.25
6	Driveway Apron	7	6	4	22	88	1.75	$ 75.00	$ 131.25
7									$3,029.50
8									
9	Concrete DB								
10		Bag Mix	Unit	Rocktec	Redimix	Hardcrete			
11		5	CY	$ 67.00	$ 66.50	$ 66.00			
12		5.5	CY	$ 69.00	$ 70.00	$ 68.80			
13		6	CY	$ 71.00	$ 71.80	$ 70.50			
14		6.5	CY	$ 73.00	$ 74.00	$ 73.50			
15		7	CY	$ 75.00	$ 76.50	$ 77.00			
16									

Cell H3 formula: =VLOOKUP(B3,ConcreteDB,3,FALSE)

Figure 6.8 Dealing with Multiple Suppliers

In the VLOOKUP formula for the unit costs, =VLOOKUP(B3,ConcreteDB,3,False), the column_index returns the value stored in column 3 of the concrete database. We wouldn't want to be required to change the formulas in H3:H6 every time we changed suppliers. We could, however, make the column_index value a variable based on a selected supplier.

The **MATCH** function can help us solve this problem. The MATCH function returns the relative position of a specified item in a list. For example, in the list Red, Green, Blue, Yellow, selecting Green would return the value 2.

The syntax, or form, of the **MATCH** function is:

MATCH(lookup_value,lookup_array,match_type)

Lookup_value is the value you use to look for a matching value in a list. **Lookup_array** is the list that contains the possible choices (suppliers in our case).
Match_type is the number -1, 0, or 1.

- If **match_type** is 1, MATCH returns the largest value that is less than or equal to lookup_value. The Lookup_array, or list, must be placed in ascending order: …1, 2, 3…, A-Z, FALSE, TRUE.
- If **match_type** is 0, MATCH finds the first value that is exactly equal to lookup_value. Lookup_array can be in any order.
- If **match_type** is -1, MATCH finds the smallest value that is greater than or equal to lookup_value. The Lookup_array must be placed in descending order: TRUE, FALSE, Z-A, …3, 2, 1….
- If **match_type** is omitted, it is assumed to be 1.

First, we need to create a cell where a supplier's name can be entered (I1, Figure 6.9).

80 Estimating with Microsoft Excel

	A	B	C	D	E	F	G	H	I
	ConcreteSuppliers		=	Bag Mix					
1	**Flatwork Detail**					Waste Factor:	5%	Supplier:	Redimix
2	Description	Bag Mix	Thick "	Width '	Length '	SF	CY	$/Unit	Total $
3	Floor Slab	5.5	4	32	58	1856	24.25	$ 69.00	$1,673.25
4	Garage Floor	5.5	4	24	28	672	8.75	$ 69.00	$ 603.75
5	Driveway	6	4	22	30	660	8.75	$ 71.00	$ 621.25
6	Driveway Apron	7	6	4	22	88	1.75	$ 75.00	$ 131.25
7									$3,029.50
8									
9	**Concrete DB**								
10	Bag Mix	Unit	Rocktec	Redimix	Hardcrete				
11	5	CY	$ 67.00	$ 66.50	$ 66.00				
12	5.5	CY	$ 69.00	$ 70.00	$ 68.80				
13	6	CY	$ 71.00	$ 71.80	$ 70.50				
14	6.5	CY	$ 73.00	$ 74.00	$ 73.50				
15	7	CY	$ 75.00	$ 76.50	$ 77.00				
16									

Figure 6.9 Setting Up to Use the Match Function

1. Name cell **I1 ConcreteSupplier**.
2. Select cells **A10:E10**.
3. Name the range **ConcreteSuppliers**.

Assume we wanted VLOOKUP to return Redimix's price for 5.5-bag concrete. Redimix's prices are found in column 4 of the concrete database. The MATCH function that will return the correct value (4) to Vlookup's column_index is:

MATCH(ConcreteSupplier,ConcreteSuppliers,0)

MATCH(lookup_value,lookup_array,match_type)

The MATCH syntax is shown below the actual function. Below are the parameters of each function:

The Lookup_value, ConcreteSupplier in I1, is Redimix.

MATCH looks for the position of Redimix in the ConcreteSuppliers Lookup_array.

To avoid errors in calculations, we want an exact match of the supplier's name (Redimix) with a value in the Lookup_array. To specify an exact match; 0 was entered as the third parameter of the Match function.

The MATCH formula that you have created can now be inserted into the column_index parameter of the Vlookup formula.

=VLOOKUP(B3,ConcreteDB,3,FALSE)

MATCH (ConcreteSupplier,ConcreteSuppliers,0)

The resulting formula is:

=VLOOKUP(B3,ConcreteDB,Match(ConcreteSupplier,ConcreteSuppliers,0),FALSE)

1. Insert this formula into cell **H3** and copy it to cells **H4:H6** (Figure 6.10).
2. Enter Hardcrete into cell **I1**.

	A	B	C	D	E	F	G	H	I	J
1	Flatwork Detail					Waste Factor:	5%	Supplier:	Redimix	
2	Description	Bag Mix	Thick "	Width '	Length '	SF	CY	$/Unit	Total $	
3	Floor Slab	5.5	4	32	58	1856	24.25	$ 70.00	$1,697.50	
4	Garage Floor	5.5	4	24	28	672	8.75	$ 70.00	$ 612.50	
5	Driveway	6	4	22	30	660	8.75	$ 71.80	$ 628.25	
6	Driveway Apron	7	6	4	22	88	1.75	$ 76.50	$ 133.88	
7									$3,072.13	
8										
9	Concrete DB									
10	Bag Mix		Unit	Rocktec	Redimix	Hardcrete				
11	5		CY	$ 67.00	$ 66.50	$ 66.00				
12	5.5		CY	$ 69.00	$ 70.00	$ 68.80				
13	6		CY	$ 71.00	$ 71.80	$ 70.50				
14	6.5		CY	$ 73.00	$ 74.00	$ 73.50				
15	7		CY	$ 75.00	$ 76.50	$ 77.00				
16										

Figure 6.10 Nesting a MATCH Function within a VLOOKUP Function

Verify that the costs change as you enter each new supplier.

List Validation

To enter in a new supplier name, you were required to manually type in the name. This process can be simplified by using **List** Validation to enter the name with point-and-click ease. To create a **List** Validation for Cell **I1**:

1. Select cells **C10:E10**.
2. Name the selected range, **ConcreteSuppliersList**.
3. Select cell **I1**.
4. Click **Data**
5. Click **Validation**.
6. In the **Allow** box select **List**.
7. In the **Source** box, enter =ConcreteSuppliersList (or **Insert/Name/Paste** and select **ConcreteSuppliersList**).

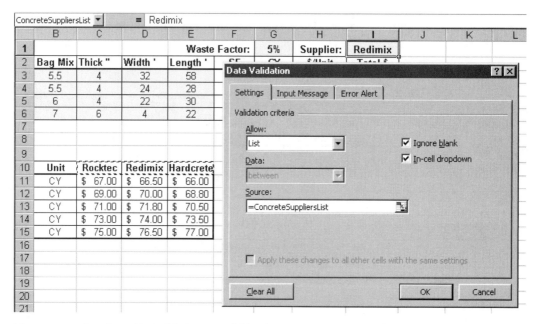

Figure 6.11 Creating a List Validation to Quickly Choose a Concrete Supplier

Click on the down-arrow next to the concrete supplier to see a drop-down-list of suppliers to choose from (Figure 6.12).

More on Formulas and Functions **83**

	A	B	C	D	E	F	G	H	I
1	**Flatwork Detail**				Waste Factor:		5%	Supplier:	Hardcrete
2	Description	Bag Mix	Thick "	Width '	Length '	SF	CY	$/Unit	Rocktec
3	Floor Slab	5.5	4	32	58	1856	24.25	$ 68.80	Redimix
4	Garage Floor	5.5	4	24	28	672	8.75	$ 68.80	Hardcrete
5	Driveway	6	4	22	30	660	8.75	$ 70.50	$ 616.88
6	Driveway Apron	7	6	4	22	88	1.75	$ 77.00	$ 134.75
7									$3,022.03
8									
9	**Concrete DB**								
10	Bag Mix	Unit	Rocktec	Redimix	Hardcrete				
11	5	CY	$ 67.00	$ 66.50	$ 66.00				
12	5.5	CY	$ 69.00	$ 70.00	$ 68.80				
13	6	CY	$ 71.00	$ 71.80	$ 70.50				
14	6.5	CY	$ 73.00	$ 74.00	$ 73.50				
15	7	CY	$ 75.00	$ 76.50	$ 77.00				
16									

Figure 6.12 Choosing from a Data Validation List

Click on Hardcrete, for example, and quickly see what the difference in total cost will be compared to Redimix.

Calculating Rebar for Footings and Foundations

Calculating the quantity of rebar for footings is not difficult, but if not calculated correctly, you may be left a bit short. Typically, footings have two continuous horizontal #4 bars; the size, grade, and number of bars will vary from job to job. Rebar comes in 20-foot lengths but should be overlapped approximately 40 bar diameters. For #4 bar, that means about 20 inches. The practical length of a bar, then, is only about 18 feet. If you account for waste, you need to subtract about 3 inches for each percent of waste.

The formula that you might use to calculate the correct quantity of bars is:

=No._of_Bars*ROUNDUP(LF/18*(1+Waste),0)

where:
No._of_Bars is the number of horizontal bars in the footings (1, 2, 3, etc.)
LF is the linear feet of footings
Waste is the percentage of waste (.03, .05, .1, etc.)

No._of_Bars, **LF**, and **Waste** are named cells that store the take-off information. The formula can also make direct reference to the cells instead of using named cells (Figure 6.13).

	A	B	C	D
	D3		=	=A3*ROUNDUP(B3/18*(1+C3),0)
1	**Footing Rebar**			
2	Number of Horiz. Bars	LF	Waste	QTY of Bars
3	2	220	3%	26
4				
5	**Footing Dowels**			
6	Spacing in Inches	LF	Waste	QTY of Dowels
7	24"	100	5%	53
8				

Figure 6.13 Formulas for Footing Rebar

The formula to calculate the quantity of needed dowels can be entered into **D7** (Figure 6.13):

=ROUNDUP(B7*12/A7*(1+C7),0)

Again, the formula could be created using named cells, for instance, Spacing, LF, and Waste, instead of using direct references in the formulas, such as, A7, B7, and C7.

=ROUNDUP(LF*12/ Spacing *(1+ Waste),0)

Custom Formatting

Notice the 24" under **Spacing in inches**. The input is 24, not 24". The spreadsheet will not accept 24" as a number but it will interpret this input as text, which will cause an error in the formula. So how did the " get in the cell? Look at the cells under Waste. These cells have been formatted for percentages. Cell A7 has been custom-formatted for inches (Figure 6.14).

More on Formulas and Functions   **85**

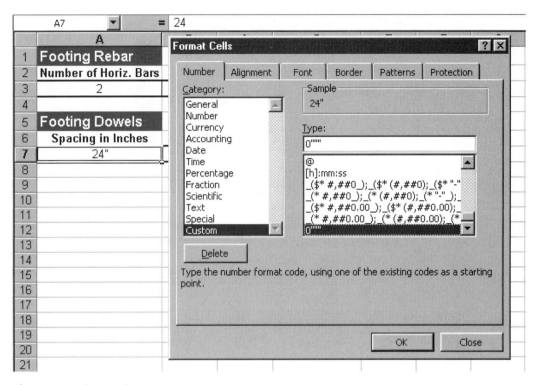

Figure 6.14 Custom Formats

1. Select **A7**.
2. Click **Format**.
3. Click **Cells**.
4. Click **Number**.
5. Click **Custom**.

In the box under Type, enter Zero, Quote, Apostrophe, Apostrophe, End Quote (0"'''").

The two apostrophes between the quotes display as a quote or an inch mark. Click **OK** to finish the formatting. Any number that is typed into the newly formatted cell will display the inch mark ("), yet Excel will still treat the data as numeric data and will be able to perform arithmetic operations on it.

Estimating Permit Fees

Estimating for permit fees can be a little frustrating. Call any municipality—*they* even hate to calculate them. The biggest problem is that you are dealing with multiple entities, each having their own method of calculating permit and plan check fees. Many municipalities have tried to standardize the process and usually base their calculations on guidelines from the major building code for the region.

Permit fees are normally based on the value of the improvements to a property. To determine the total valuation of improvements, multiply the square feet by the valuation rate for the specific type of construction (i.e., main floor, garage, deck, etc.). The permit fee is then calculated using a graduated fee structure based on the total valuation of the construction. Many municipalities determine their valuation rates based on the Building Valuation Data average building costs, which are published by *Building Standards*, a national standard that has adjustment factors for individual states.

Suppose you wanted to build a 1,700-square foot (SF) slab-on-grade residence in Salem. The plans call for a 550-square foot garage and 140-square foot deck (Figure 6.15).

A1		=	=April Parkway	
	A	B	C	D
1	$ 749,000			
2				

Figure 6.15 Calculating Construction Valuation

The valuation rate for each type of construction for Salem is listed. The main floor area is valued at $66.00/SF; garage SF floor area is valued at $23.80/SF, and so forth.

Multiply the SF areas by the rate for each type of construction (=B4*C4).
The total valuation for the job is the sum of the cost extensions ($126,690.00). So how can we determine the fees using Excel?

1. Select cell **D11**
2. Click the **AutoSum** button.

Most municipalities use the standard permit fee table from the codebook to calculate the permit and plan check fees. Figure 6.16 shows the permit fee structure from the 1997 Uniform Building Code (UBC). Other codes will have fee structures similar to this one. Once the building valuation has been determined, locate the corresponding valuation in columns A and B of the fee structure table. Our valuation ($126,690.00) falls between 100,001 and 500,000 (row 5).

	A	B	C	D	E	F
1	**Permit Fee Structure UBC - 1997**					
2	For Property Improvement Valuation From:	To:	Base Fee	For the 1st	Plus an additional	For Each __ Above Base Fee
3	$ 1,000,001	$1,000,000,000	$5,608.75	$ 1,000,000	3.65	$ 1,000.00
4	$ 500,001	$ 1,000,000	$3,233.75	$ 500,000	4.75	$ 1,000.00
5	$ 100,001	$ 500,000	$ 993.75	$ 100,000	5.60	$ 1,000.00
6	$ 50,001	$ 100,000	$ 643.75	$ 50,000	7.00	$ 1,000.00
7	$ 25,001	$ 50,000	$ 391.25	$ 25,000	10.10	$ 1,000.00
8	$ 2,001	$ 25,000	$ 69.25	$ 2,000	14.00	$ 1,000.00
9	$ 501	$ 2,000	$ 23.50	$ 500	3.05	$ 100.00
10	$ 1	$ 500	$ 23.50	$ -	0	$ -
11						

Figure 6.16 UBC Building Permit Fee Structure

For the first $100,000, the building permit fee will be $993.75. For each additional $1,000.00 of Valuation (rounded up), we can move over in row 5 and see we must add $5.60. $126,690–$100,000 = $26,690. $26,690 / $1,000 = 26.7, or rounded up, 27 x $5.60 = $151.20. The total permit fee is $993.75 + $151.20 = $1,144.95. A formula that will calculate the permit fee for the $100,001 to $500,000 range is:

= 993.75+(ROUNDUP((Valu-100000)/1000,0)*5.6)

where **Valu** is the total calculated valuation ($126,690.00).

	D14		=0.65*D13	
	A	B	C	D
1	**Permits**			
2	**Salem**	SF	Valuation/SF	Total Valuation
3				
4	Main Floor SF	1700	66.00	112,200.00
5	2nd Floor SF		66.00	0.00
6	Finished Basement SF		17.50	0.00
7	**Total Finished SF**	1700		
8	Unfinished Basement SF		12.50	0.00
9	Garage SF	550	23.80	13,090.00
10	Deck	140	10.00	1,400.00
11	**Total Valuation**			**126,690.00**
12				
13	Permit			$ 1,144.95
14	Plan Check			744.22
15	State Fee			11.45
16	**SubTotal**			$ **1,900.62**
17				

Figure 6.17 Calculating Permit and Associated Fees.

The UBC stipulates that the plan check fee is 65 percent of the total building permit fee (The calculated value, $744.87, will vary slightly depending on the precision of rounding—$744.22). The plan check fee charged by some municipalities is only 40 percent. The percentage can vary from municipality to municipality. Some states will also add a one percent or two percent tax to the permit fee.

The permit fee formula = 993.75+(ROUNDUP((Valu-100000)/1000,0)*5.6) is fine if you only do construction valued between $100,001 to $500,000. The problem is developing a formula to calculate the permit fee for *any* valuation. If the total valuation were greater than $1,000,000.00, then you would use one formula. However, if the valuation were greater than $500,000.00, then you would use another formula, etc. Table 6.1 displays the formulas that would be needed to calculate the permit fee for any valuation range.

Table 6.1 Formulas to Calculate the Permit Fee For Different Valuations

```
IF(Valuation>1000000,(5608.75+(ROUNDUP((Valuation-1000000)/1000,0)*3.65))
IF(Valuation>500000,(3233.75+(ROUNDUP((Valuation-500000)/1000,0)*4.75))
IF(Valuation>100000,(993.75+(ROUNDUP((Valuation-100000)/1000,0)*5.6))
IF(Valuation>50000,(643.75+(ROUNDUP((Valuation-50000)/1000,0)*7))
IF(Valuation>25000,(391.25+(ROUNDUP((Valuation-25000)/1000,0)*10.1))
IF(Valuation>2000,(69.25+(ROUNDUP((Valuation-2000)/1000,0)*14))
IF(Valuation>500,(23.5+(ROUNDUP((Valuation-500)/1000,0)*3.05))
23.5
```

Note: Numbers that are entered in formulas should be entered without commas, periods, or other formatting symbols. If you enter a number with a comma, Excel reads it as text. Enter only the number and format the cell if needed.

Nest these IF formulas together into one formula, and it will correctly calculate any valuation. Nesting formulas together can quickly get confusing. If you break it into parts as in Table 6.1, it is a little easier to keep organized. The maximum number of nested IF statements is 7. We have nested the maximum number of IF statements in our formula:

```
=IF(Valuation>1000000,(5608.75+(ROUNDUP((Valuation-1000000)/1000,0)*3.65)),
IF(Valuation>500000,(3233.75+(ROUNDUP((Valuation-500000)/1000,0)*4.75)),
IF(Valuation>100000,(993.75+(ROUNDUP((Valuation-100000)/1000,0)*5.6)),
IF(Valuation>50000,(643.75+(ROUNDUP((Valuation-50000)/1000,0)*7)),
IF(Valuation>25000,(391.25+(ROUNDUP((Valuation-25000)/1000,0)*10.1)),
IF(Valuation>2000,(69.25+(ROUNDUP((Valuation-2000)/1000,0)*14)),
IF(Valuation>500,(23.5+(ROUNDUP((Valuation-500)/100,0)*3.05)),23.5)))))))
```

Protecting Cells from Accidental Erasure

It is a tedious task to make sure that the parentheses match up in their proper locations. If one comma or parenthesis is missing or out of place, then the formula will crash or return a wrong answer. You would want to protect this cell from accidental deletion. Be sure that the Total Valuation cell is protected. To protect cells from accidentally overwriting them:

1. Select the **cell** or cells that you want to protect.
2. Click **Format**.
3. Click **Cells**.
4. Click **Protection**.
5. Click the box next to **Locked** so that it is checked.

Once the cells you want protected are locked, you must protect the sheet. To protect the sheet:

1. Click **Tools** on the **Menu** bar.
2. Click **Protection**.
3. Click **Protect Sheet**.

Excel prompts you to enter a password for your protected sheet. If you want to only protect against accidentally overwriting a cell, you won't want to enter a password. If you do enter a password, be sure to write it down and don't lose it or forget it because you will not be able to make further changes without the password.

Working with Variables

It is better to keep variables (e.g., 1000000, 5608.75, 3.65, etc.) out of the formula itself. When building codes change, revisions can be made directly to the Permit Fee table and the Permit Fee formula is automatically updated through its references to individual cells in the Permit Fee table (see formula bar in figure 6.18).

B14		=IF(Valuation>B4,(C3+(ROUNDUP((Valuation-B4)/1000,0)*E3)),IF(Valuation>B5,(C4+(ROUNDUP((Valuation-B5)/1000,0)*E4)),IF(Valuation>B6,(C5+(ROUNDUP((Valuation-B6)/1000,0)*E5)),IF(Valuation>B7,(C6+(ROUNDUP((Valuation-B7)/1000,0)*E6)),IF(Valuation>B8,(C7+(ROUNDUP((Valuation-B8)/1000,0)*E7)),IF(Valuation>B9,(C8+(ROUNDUP((Valuation-B9)/1000,0)*E8)),IF(Valuation>B10,(C9+(ROUNDUP((Valuation-B10)/100,0)*E9)),C10)))))))
	A	

	A						
1	Permit Fee Structure UBC - 1997						
2	For Property Improvement Valuation From:		To:	Base Fee	For the 1st	Plus an additional	__ Above Base Fee
3	$	1,000,001	$ 1,000,000,000	$ 5,608.75	$ 1,000,000	3.65	$ 1,000.00
4	$	500,001	$ 1,000,000	$ 3,233.75	$ 500,000	4.75	$ 1,000.00
5	$	100,001	$ 500,000	$ 993.75	$ 100,000	5.60	$ 1,000.00
6	$	50,001	$ 100,000	$ 643.75	$ 50,000	7.00	$ 1,000.00
7	$	25,001	$ 50,000	$ 391.25	$ 25,000	10.10	$ 1,000.00
8	$	2,001	$ 25,000	$ 69.25	$ 2,000	14.00	$ 1,000.00
9	$	501	$ 2,000	$ 23.50	$ 500	3.05	$ 100.00
10	$	1	$ 500	$ 23.50	$ -	0	$ -
11							
12							
13	Valuation		Permit Fee				
14	$	126,690.00	$1,144.95				
15							

Figure 6.18 Formula References Cells in the Permit Fee Table (PermitUBC Sheet)

Connection and Impact Fees

Connection and impact fees vary depending on the municipality in which you build. These fees are normally revised each year. To set up an estimating sheet for connection and impact fees, start by creating a database of fees for each municipality in which you build (Figure 6.19).

C16		=IF(WaterLateralSize=" 3/4 inch",C17,IF(WaterLateralSize=" 1 inch",C18,"Lateral size?"))					
	A	B	C	D	E	F	G
11	**Fees**		Salem	Payson	Mapleton	Woodland Hills	Spanish Fork
12	Description		423-2770	465-5214	489-5655	423-1962	798-5080
13	Sewer Connection		155.00	175.00	196.00		
14	Sewer Impact		960.00	850.75	2,814.07		1,326.00
15	Storm Water						416.00
16	Water Connection		223.00	100.00	0.00	2,000.00	0.00
17	3/4 inch		184.00	50.00		1,600.00	
18	1 inch		223.00	100.00		2,000.00	
19	Water Impact		2,500.00	2,249.97	1,238.50		1,069.00
20	Meter Set Only					160.00	
21	Pressurized Irrigation			125.00			
22	Electrical Connection		500.00	250.00			
23	Electrical Impact		1,108.50	984.13			442.00
24	Parks & Recreation Impact		1,000.00	488.96	1,807.36		1,100.00
25	Safety				842.75		
26	Bonds				1,000.00		
27	Temporary Power			200.00			
28	Construction Water Use		50.00				
29	Permit Fee			22.00			
30	Electrical Permit			20.00			
31	New service up to 200 amp			3.00			
32	Plumbing Permit			20.00			
33	# of fixtures			7.00			
34	Water heater and vent			7.00			
35	Gas line - up to 5 outlets			5.00			
36	Mechanical Permit			22.00			
37	New Furnace			16.25			
38	Appliance Vents			6.50			
39	Miscellaneous					2,900.00	
40	**Total Fees**		6,496.50	5,552.56	7,898.68	5,060.00	4,353.00

Figure 6.19 Connection and Impact Fee Database

On the detail sheet (Figure 6.20) create a list of fee descriptions that match those in the connection fee database. You will need to write a formula that will return the unit prices from the connection fee database. This can be accomplished using the VLOOKUP function. The unit prices that VLOOKUP returns should also be based on the municipality that has jurisdiction over the construction. Use the MATCH function to return fee amounts based on the municipality (cell A2, Figure 6.15).

	A	B	C	D	E
	B20		=VLOOKUP(A20,FeesDB,MATCH(City,CitiesList,0),FALSE)		
18	**Fees**				
19	Salem				
20	Sewer Connection	155.00			
21	Sewer Impact	960.00			
22	Storm Water	0.00	Lateral Size:		
23	Water Connection	223.00	**1 inch**		
24	3/4 inch	184.00			
25	1 inch	223.00			
26	Water Impact	2,500.00			
27	Meter Set Only	0.00			
28	Pressurized Irrigation	0.00	Subdivision Lot?		
29	Electrical Connection	500.00	**Yes**		
30	Electrical Impact	1,108.50			
31	Parks & Recreation Impact	1,000.00			
32	Safety	0.00			
33	Bonds	0.00			
34	Temporary Power	0.00			
35	Construction Water Use	50.00			
36	Permit Fee	0.00			
37	Electrical Permit	0.00			
38	New service up to 200 amp	0.00			
39	Plumbing Permit	0.00			
40	# of fixtures	0.00			
41	Water heater and vent	0.00			
42	Gas line - up to 5 outlets	0.00			
43	Mechanical Permit	0.00			
44	New Furnace	0.00			
45	Appliance Vents	0.00			
46	Miscellaneous	0.00			
47	Salem				
48	**Total Fees**	**6,496.50**			
49					

Figure 6.20 Connection and Impact Fee Detail

Notice that the water connection fee depends on the size of the water service. Select the size of water service and the price is automatically adjusted (see Edit Bar in Figure 6.19). Other fees may depend on specific lot or municipality variables. These can also be addressed in the same manner as the water connection fees. As a visual check that you have selected the correct city, A19 and A47 are linked to the cell where the city is selected (from a list validation) or typed in. To link these cells:

1. Select cell **A19**
2. Enter =City
3. Repeat this process for cell **A47**.

Adding Comments to Cells

Occasionally, you may wish to give additional clarification or special instructions about information in a specific cell (cell **C24**, Figure 6.21). As you pass, or pause the cursor over a cell that contains a comment, the comment will become visible automatically.

	A	B	C	D	E	
						Wood
11	Fees		Salem	Payson	Mapleton	Hills
12	Description		423-2770	465-5214	489-5655	423-19
13	Sewer Connection		155.00	175.00	196.00	
14	Sewer Impact		960.00	850.75	2,814.07	
15	Storm Water					
16	Water Connection		223.00	100.00	0.00	2,
17	3/4 inch		184.00	50.00		1,
18	1 inch		223.00	100.00		2,
19	Water Impact		2,500.00	2,249.97	1,238.50	
20	Meter Set Only					
21	Pressurized Irrigation			125.00		
22	Electrical Connection		500.00	250.00		
23	Electrical Impact		1,108.50			
24	Parks & Recreation Impact		1,000.00	South Valley View Subdivision plats A-J exempt from this impact fee.	36	
25	Safety				75	
26	Bonds				00	
27	Temporary Power			200.00		
28	Construction Water Use		50.00			

Figure 6.21 Adding Comments to a Worksheet

To insert a comment into a cell:

1. Select the cell to which you want to add a comment.
2. Click **Insert**
3. Click **Comment**. A comment box appears next to the cell. The comment box is in Edit mode and is ready for you to input notes or instructions for the user.
4. Enter the text for the comment.
5. After a comment has been created, it can be edited or deleted by right clicking over the cell and clicking **Edit Comment** or **Delete Comment** from the drop-down menu.

A sample comment was created in cell workbook. To cause the comment to remain visible even when the cursor is not over the cell that contains the comment:

1. Right-click on **C24** of the **PermitFeesDB** sheet in Chapter 6B Formulas.xls.
2. Click **Show Comment**.

To hide the comment so that it once again appears only when the cursor passes over the cell:

1. Right-click cell **C24**.
2. Click **Hide Comment**.

Roofing

Builders can spend a lot of time estimating the roofing for a building. I have asked many builders and remodelers how they estimate roofing. Most of the time I get answers like, I pull out the elevation plans, measure the area of each slope of the roof, then add them together, or I climb on the roof and within 30 minutes or so, I can get the exact measurements of the roof. The problem, then, is that the person has already spent half an hour driving to the job and will drive another half hour to the office and then spend another half hour calculating quantities and making a material list—that's two hours!

Seasoned framers can cut rafters for an entire hip roof without getting on the roof. They make calculations for ridge cuts and bird's mouths and cut the rafters on the ground. The ridge is then lifted into place, the rafters are installed and the roof is trimmed out. I see many framers set the ridge and then climb up and down the ladder a number of times to measure, mark, cut, check, measure again, mark, cut, check, etc. before they discover a rafter that they can use for a pattern. It gets worse for hips, jacks, and valleys.

The same principles that allowed the seasoned framers to calculate their rafters on the ground can be used to calculate the squares of shingles needed for the roof. People are amazed when they learn how fast and accurately they can estimate roofing.

Figure 6.22 is a plan view of a typical hip roof. The dimensions in the figure are to the fascia. From this view, the roof appears flat. If we were to calculate the shingles needed for this flat roof, we would determine the area of the roof and add for starter shingles (since there would be no hip and ridge shingles on a flat roof).

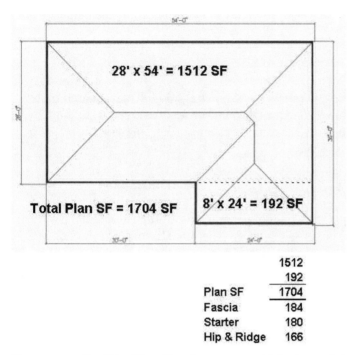

	1512
	192
Plan SF	1704
Fascia	184
Starter	180
Hip & Ridge	166

Figure 6.22 Roof Plan View (Drawing courtesy of John Jones, Softplan)

For ease of calculation, the plan-view area of the roof is divided into two rectangles, one 28 feet by 54 feet and the other 8 feet by 24 feet. The total flat SF area is the sum of 1,512 SF and 192 SF, or 1,704 SF. It will take approximately 17 one-third squares of shingles—not including starter strip or waste to cover this roof (if it were flat).

This roof has a 6/12 slope, that is, for every 12 inches of run (horizontal distance) the roof rises 6 inches. If we know the rise and the run of a roof slope, we can determine the length of the sloped section (Figure 6.23).

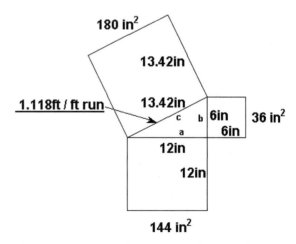

Figure 6.23 Conversion Factor for the Roof Slope

The basic relationship of the sides of a right triangle can be expressed by the Pythagorean formula:

$$a^2 + b^2 = c^2$$

Meaning: The squared area of side a (12 inches by 12 inches = 144 inches squared) plus the squared area of side b (6 inches by 6 inches = 36 inches squared) is equal to the squared area of c (180 inches squared). Take the square root (the length of one side of a square) of 180 to find the length of side c.

$$c = \sqrt{a^2 + b^2} = \sqrt{12''^2 + 6''^2} = \sqrt{180''} = 13.42''$$

For the 6/12-sloped roof, this equation says that for every 1 ft of run (horizontal distance) the length along the slope of the roof is 13.42 inches. The slope factor, 13.42 inches, if converted to feet, equals 1.118 feet = (13.42 in./12 in./ft.).

The slope factor (1.18) can be used to convert the flat SF of an area to a sloped SF area. Our example roof has 1,704 flat SF. Multiply the flat SF by the slope factor to find the SF of sloped roof area.

$$1704 \text{ ft}^2 \times 1.118 = 1905 \text{ ft}^2$$

Not counting starter strip, hip and ridge shingles, or waste, we would need to order 19 one-third squares of shingles to cover this 6/12-sloped roof. You can use this procedure to calculate roof sheathing also. Adjust the waste factor for roof sheathing to allow for cuts and overbuilds.

A formula to calculate the slope factor for any slope is shown in the Formula bar in Figure 6.24. **SQRT** will take the square root of whatever follows in the parentheses. Cell **B5** has been given the name **RoofSlope**. The ^ symbol is the power sign; the **RoofSlope** is being raised to the power of two (two is being squared).

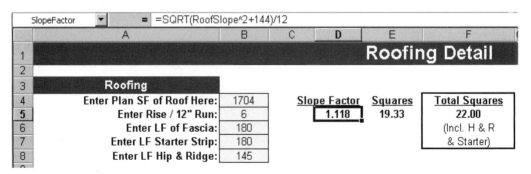

Figure 6.24 Slope Factor Formula

Once the slope factor has been calculated, the quantity of squares can easily be figured. The formula for the squares of shingles is shown in the Formula bar in Figure 6.25. The basic formula is:

<ds>=B4*D5/100.

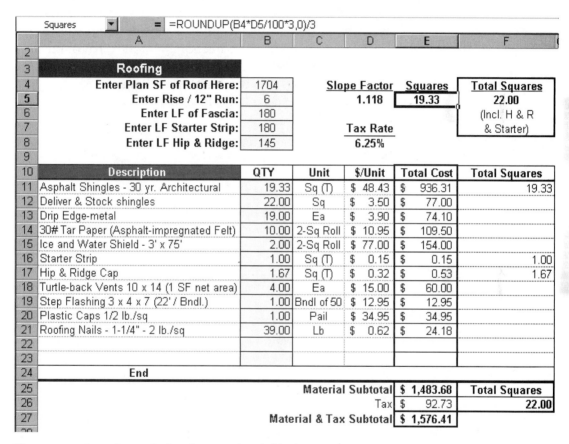

Figure 6.25 Formula to Calculate the Quantity of Shingles

The ROUNDUP function was added to round the quantity up to the nearest bundle (one-third square).

=ROUNDUP(B4*D5/100*3,0)/3.

The total quantity of shingles, after adding starter strip hip and ridge is shown in cell **F5**.

Starter Strip

To calculate starter strip, take the LF of the eaves only (not the rakes or gabled ends) and divide by 240. One square of shingles covers approximately 240 linear feet of starter. One square of shingles covers approximately 100 LF of hip and ridge.

> **TIP**
>
> When using the **ROUNDUP** function to round up to fractions of a whole, multiply the number by the fractional amount before rounding up to the nearest whole number then divide the results by the fractional amount. To round up to the nearest fourth, multiply the number by four, round up, and then divide by four.
>
> =ROUNDUP(Number*4,0)/4.
>
> To round a number up to the nearest sixth, multiply the number by six, round up, then divide by six, etc.

Roofing Labor

The labor rate per square will depend on the slope of the roof. Figure 6.26 shows a typical roofing labor database. For example, the labor rate for an asphalt shingle 30-yr. architectural roof with either a 5/12 or a 6/12 slope would be $43.00.

The Vlookup formula (Figure 6.27) retrieves the value for the roofing labor. The MATCH function locates the column in the database with the correct slope factor. If the match_type in the MATCH function is omitted, the default value is assumed to be 1. If match_type is 1, MATCH finds the largest value that is less than or equal to lookup_value. Lookup_array must be placed in ascending order (e.g., blank, 4, 6, 8, 10, 12—LaborSlopeList, Figure 6.26).

	B	C	D	E	F	G
33	**Labor**			**Slope**		
34		4	6	8	10	12
35	Asphalt Shingles - 25 yr. 3 Tab	35	40	50	55	60
36	Asphalt Shingles - 30 yr. 3 Tab	35	40	50	55	60
37	Asphalt Shingles - 30 yr. Architectural	38	43	53	58	65
38	Cedar Shakes #1 medium handsplits	60	65	75	85	90
39	Cedar Shingles #1	60	65	75	85	90
40	Metal Roofing	70	70	80	85	90
41	Eagle Tile	60	65	75	85	90
42						
43	**Labor & Material**			**Slope**		
44		4	6	8	10	12
45	Asphalt Shingles - 25 yr. 3 Tab	100	105	115	125	130
46	Asphalt Shingles - 30 yr. 3 Tab	105	110	120	130	135
47	Asphalt Shingles - 30 yr. Architectural	125	130	135	140	145
48	Cedar Shakes #1 medium handsplits	190	195	210	220	230
49	Cedar Shingles #1	220	225	240	250	260
50	Metal Roofing	170	175	185	195	200
51	Eagle Tile	185.00	190	200	210	215
52						

Figure 6.26 Roofing Labor Database

Sub-Bid Methods

Figure 6.27 shows an alternative method of estimating that can be used instead of the detail method that we have just discussed. The sub-bid method is faster but may be not

	A	B	C	D	E	F
25				Material Subtotal	$ 1,483.68	**Total Squares**
26				Tax	$ 92.73	22.00
27				Material & Tax Subtotal	$ 1,576.41	
28						
29	**Roofing Labor**					
30	Description	Qty	Unit	$/Unit	Total Cost	
31	Asphalt Shingles - 30 yr. Architectural	22.00	Sq	$ 43.00	$ 946.00	
32						
33						
34	**Sub-Bid Roofing Labor & Material**					
35	Description	Qty	Unit	$/Unit	Total Cost	
36		22.00	Sq		$ -	
37						
38	Total Material and Labor				$ 2,522.41	

Figure 6.27 Roofing Labor and Roofing Sub-Bid

quite as accurate. It can be helpful in performing what-if scenarios. What if we put 25-year shingles on the roof or wanted to know a quick cost for wood shakes? Figures 6.28 and 6.29 show the results of these two queries.

	A	B	C	D	E	F
25				Material Subtotal	$ 1,483.68	Total Squares
26				Tax	$ 92.73	22.00
27				Material & Tax Subtotal	$ 1,576.41	
28						
29	**Roofing Labor**					
30	Description	Qty	Unit	$/Unit	Total Cost	
31	Asphalt Shingles - 30 yr. Architectural	22.00	Sq	$ 43.00	$ 946.00	
32						
33						
34	**Sub-Bid Roofing Labor & Material**					
35	Description	Qty	Unit	$/Unit	Total Cost	
36	Asphalt Shingles - 25 yr. 3 Tab	22.00	Sq	$ 105.00	$ 2,310.00	
37						
38	Total Bid Price				$ 2,310.00	

Figure 6.28 25-Year Asphalt Shingle Sub-Bid

= =IF(A36="",E27+E31,E36)

	A	B	C	D	E	F
25				Material Subtotal	$ 1,483.68	Total Squares
26				Tax	$ 92.73	22.00
27				Material & Tax Subtotal	$ 1,576.41	
28						
29	**Roofing Labor**					
30	Description	Qty	Unit	$/Unit	Total Cost	
31	Asphalt Shingles - 30 yr. Architectural	22.00	Sq	$ 43.00	$ 946.00	
32						
33						
34	**Sub-Bid Roofing Labor & Material**					
35	Description	Qty	Unit	$/Unit	Total Cost	
36	Cedar Shakes #1 medium handsplits	22.00	Sq	$ 195.00	$ 4,290.00	
37						
38	Total Bid Price				$ 4,290.00	

Figure 6.29 Cedar Shakes Sub-Bid

Notice in cell E38 that when A36 has an entry, the total reflects the sub-bid costs. If cell A36 is empty (""), then the detail material and labor prices are added together and displayed as the total. The description for the total cost (E38) also depends upon whether or not there is input into cell A36. IF functions control the input into both cells (A38 and E38). The formula for cell **A38** is:

=IF(A36="","Total Material and Labor","Total Bid Price")

and the formula for cell **E38** is:

=IF(A36="",E27+E31,E36)

What-If Scenarios

The completed Roofing Detail sheet allows the user to quickly and accurately create detailed take-off and cost estimates or sub-bid cost estimates. Your customer asks, What if we change to an 8/12 slope roof, how much will it cost? Now you no longer have to make the educated guess or have the customer wait until you can redo the complete roofing estimate. Simply change RoofSlope to eight and all quantities and costs are automatically updated.

Summary

Upon completion of this chapter, you will have created detail and database sheets to help in estimating concrete, permits and fees, and roofing (See Chapter **6B Formulas.xls** on the CD that accompanies this book). The principles you have been taught can be applied to estimate other work breakdown items also.

Chapter 7 will focus on a couple of areas, such as construction loan interest and builder's margin, that can be difficult, at best, to estimate. I find that many builders lose more money by miscalculating their profit and overhead than they do by making errors in calculating quantities. Chapter 7 discusses these two areas and shows you how to use Excel to solve these problems.

Loan Interest and Builder's Margin

CHAPTER 7

IN THIS CHAPTER

- Find out about methods of calculating construction loan interest and builder's margin.
- Learn how to use Excel to instantly calculate accurate loan interest and builder's margin.
- Quickly analyze what-if scenarios that will help you price jobs so that they are profitable.

Builders and remodelers focus much of their estimating efforts on making their estimates accurate to within one or two percent of their actual costs, yet some of the most costly mistakes that I see have to do with not correctly calculating builder's margin (profit and company overhead) and loan interest. I consistently see builders who only make about half of the profit that they thought they would, and it was because they calculated their builder's profit margin incorrectly in their estimates.

This chapter will walk you through and explain methods of calculating construction loan interest and builder's margin so that you will know exactly what your real profit is. You will learn how to use Excel to instantly calculate the correct numbers.

Construction Loan Interest

Among the typical costs associated with construction loans are:

- Origination fees
- Loan interest
- Appraisal fees
- Inspection fees
- Title insurance

- Closing fees
- Miscellaneous fees

Origination Fees

Origination fees vary and may be between one and one-and-a-half percent of the amount of the construction loan. Suppose a home you want to build appraises for $250,000. If the loan-to-value ratio (LTV) is 80 percent and the origination fee rate is one percent, a formula that would calculate the origination fee amount and could be entered into cell D16 (Figure 7.1) is:

=D12*G10

	C	D	E	F	G
8					
9	**Construction Loan**				
10	Appraised Value:	$ 250,000.00		Origination Fee:	1.00%
11	LTV:	80%		Interest Rate:	9.50%
12	Loan Amount:	$ 200,000.00		Term: (Months)	6
13					
14	Description				
15	**Other Fees:**				
16	Origination Fee	$ 2,000.00			
17	Interest Reserve	$ 4,750.00			
18	Appraisal	$ 350.00			
19	Inspection Fee	$ 150.00			
20	Title Insurance	$ 455.00			
21	Closing Fees	$ 75.00			
22	Other Fees				
23	Total Construction Loan Costs	$ 7,780.00			
24					

Figure 7.1 Construction Loan Interest and Fees

If the cells that contain the appraised value, LTV ratio, loan amount, origination fee, and term for the loan have been named, a formula that communicates better might be written:

=Loan_Amount*Origination_Fee

Construction Loan Interest

Estimating the interest reserve is a little more complicated. The interest reserve is the amount of money set aside to cover interest charges for money borrowed during the course of construction. Look at the representation of the cumulated costs of construction over the duration of the project (Figure 7.2). This figure shows a typical construction S-curve.

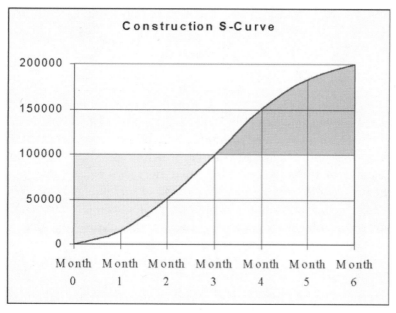

Figure 7.2 Typical Construction S-Curve

The area under the curve represents the amount of money borrowed during the course of construction. The graph is cumulative, meaning that the amount of money drawn from bank for the first month is added to the amount that is drawn for the second month and so forth. By the end of the sixth month, the total amount of the loan ($200,000) has been used.

The graph of the construction S-curve shown in Figure 7.2 was generated from the following data:

Month	Monthly Draw	Cumulative
Month 0		
Month 1	$15,100	$ 15,100
Month 2	$35,000	$ 50,100
Month 3	$49,300	$ 99,400
Month 4	$51,500	$150,900
Month 5	$32,600	$183,500
Month 6	$16,500	$200,000

At the end of Month 1, interest owing is based only on the $15,100 used. The simple interest amount for the first month is calculated by multiplying the monthly interest rate (.095/12) by the outstanding amount of the loan ($15,100), which equals $119.54. For the fifth month, interest is required on the cumulated $183,500 of borrowed money. The amount of interest for the fifth month would be equal to $1,452.71 (.095/12*183,500). We could use this procedure for each of the six months. Adding together the interest for each of the months would indicate the total interest reserve needed for the construction loan.

Another quicker and possibly just as accurate method to determine the total interest reserve is denoted by the formula:

= Average Loan Amount * Monthly Interest Rate * Term in Months

Through a leveling process of averaging the monthly loan amounts, we can determine the average monthly interest required. Look again at Figure 7.2. If the shaded area in the upper right part under the S-curve were placed in the trough (the shaded area above the lower part of the S-curve), it would be nearly a perfect rectangle (Figure 7.3). Each month's interest would be half of the loan amount multiplied by the monthly interest rate and could be written:

Average Monthly Interest Amount = Loan_Amount/2 * Interest_Rate/12

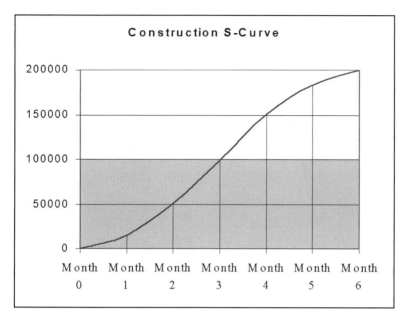

Figure 7.3 Averaging the Monthly Loan Amounts

Since the average monthly interest would be the same throughout the duration of construction, the total interest reserve would be equal to the average monthly interest amount multiplied by the term of construction (in months).
In cell D17, Figure 7.1, enter the formula:

= Loan_Amount/2 * Interest_Rate/12 * Term_Months

where **Loan_Amount** is the name given to cell D12, Figure 7.1.
Interest_Rate, the annual interest rate, is the name given to cell G11, Figure 7.1.
Term_Months is the name given to cell G12, Figure 7.1.

Builder's Margin: Profit and Company Overhead

Profit and company overhead must be added to the direct costs of construction to determine the sales price. Unfortunately, many builders incorrectly calculate the builder's margin and as a result end up with only about half of their expected profit. These builders calculate profit and company overhead based on a multiplier and not on markup.

A company's yearly gross sales can be broken down into direct costs, company overhead costs, and profit (Figure 7.4). *Direct costs* are hard costs (such as material, labor, and

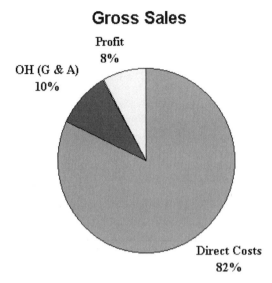

Figure 7.4 Breakout of Direct Costs, Company Overhead Costs (G&A), and Profit

equipment) plus project overhead costs (soft costs such as permits, fees, and temporary utilities). Any cost that can be attributed to a specific job is a direct cost. *Company overhead costs,* often referred to as General and Administrative (G&A) costs are the costs of doing business and include costs such as officer salaries, advertising, liability insurance, main office expenses, etc. *Profit* is the incentive for the risk that builders and remodelers take to complete a project; it is the money that is left over after all the bills are paid.

Profit and company overhead are often expressed as percentages of gross sales. Suppose that sales for the past year were good; company overhead was 10 percent of sales and profit was 8 percent of sales. Sales for the coming year looked to be about the same as the previous year. You have estimated direct costs for an upcoming project at $100,000.00. What is the sales price that you should charge your customer to make your profit and company overhead percentages? Take a minute and write your answer.

Did you figure $118,000.00, or even $118,800.00? If you did, your gross profit will be about half of what you expected. Looking at Figure 7.4, we can see that direct costs are 82 percent of the total sales price; written as a formula, it looks like:

Direct Costs = .82 * Total Sales Price

To calculate the total sales price, divide both sides by .82:

$$\frac{\text{Direct Costs}}{.82} = \frac{.82 * \text{Total Sales Price}}{.82}$$

$$\text{Total Sales Price} = \frac{\text{Direct Costs}}{.82} = \$121,951.22$$

The correct sales price should be $121,951.22. Knowing this, take ten percent of $121,951.22 to find the correct overhead and eight percent to find the right profit:

$$\text{Overhead} = \$121,951.22 * .1 = \$12,195.12$$

$$\text{Profit} = \$121,951.22 * .08 = \$9,756.10$$

Suppose the sales price had been $118,000. To find the profit, subtract the direct costs from the sales price. Subtract the correct overhead costs of $12,195, which results in $5,805 for profit. Whether or not the correct overhead was *estimated,* the correct overhead has to be paid, leaving the surplus for profit.

Sales Price	$118,000
−Direct Costs	$100,000
	$18,000
−Overhead	$12,195
Profit	$5,805

$$\text{Profit Percentage} = \frac{\$5,805}{\$118,000} = 4.9\%$$

Remember that the target profit was eight percent. The problem comes when profit is calculated as a percentage of direct costs. Figure 7.4 graphically shows that eight percent of direct costs (part of the pie) is a smaller number than eight percent of the sales price (the whole pie).

To find a multiplier that correctly calculates the sales price, divide one by one minus the margin.

$$\text{Multiplier} = \frac{1}{1 - \text{Margin}}$$

For a markup of ten percent, a multiplier of 1.11 is used.

$$\frac{1}{1-.10} = \frac{1}{.90} = 1.11$$

For a markup of 15 percent, a multiplier of 1.18 is used.
For a markup of 20 percent, a multiplier of 1.25 is used.

Circular References

A circular reference is created when a cell's value depends on the value of another cell and the value of the other cell depends on the value of the first cell. When a circular reference is made in error, it can be a real problem and Excel will notify you that it cannot calculate a circular reference.

There are times, however, when a circular reference can be very useful, for example, when calculating the sales price (Figure 7.5). The sales price depends upon the direct costs, overhead, and profit; the sales price is the sum of direct costs, overhead, and profit. Overhead and profit depend on the sales price; overhead is a percentage of the sales price (ten percent), and profit is also a percentage of the sales price (eight percent).

Direct Costs	100000
Overhead	=0.1*Sales_Price
Profit	=0.08*Sales_Price
Sales Price	=SUM(C5:C7)

Figure 7.5 Formulas that Form Circular References

Figure 7.5 shows the spreadsheet with the formulas view turned on. Toggle the formulas view (type CTRL+` again) and the results of the formulas are shown (Figure 7.6).

Direct Costs	$	100,000.00
Overhead	$	12,195.12
Profit	$	9,756.10
Sales Price	$	121,951.22

Figure 7.6 The Results of Formulas Using Circular References

Excel has the ability to solve circular references through a process called iteration. To enable Excel to calculate circular references and not give an error message, click on **Tools/Options** and select the **Calculation** tab (Figure 7.7), then click in the **Iteration** check box.

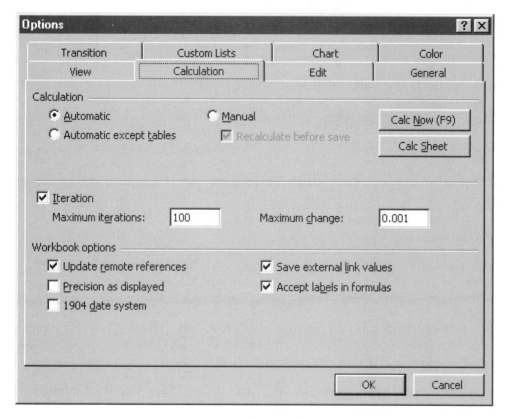

Figure 7.7 Activating Iteration to Enable Calculation of Circular References

If a circular reference was accidentally created, you may need help tracing the dependant and precedent references. You can get help through the Circular Reference Toolbar (Figure 7.8).

> **NOTE**
>
> When using the Circular Reference tools, **Iteration** should be turned off on the **Calculation** tab of the **Tools/Options Menu**.

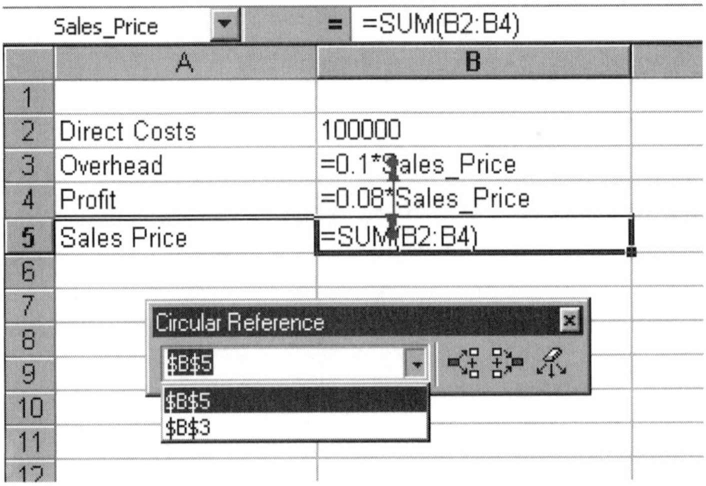

Figure 7.8 Using the Circular Reference Toolbar to Trace Errors

To open the **Circular Reference** toolbar, click **Tools/Customize** and click the box next to **Circular Reference**.

On the **Circular Reference toolbar**, click the first cell in the **Navigate Circular Reference** box. Review the formulas in the cells listed in the **Navigate Circular Reference** box and make any corrections until the status bar no longer displays the word circular.

The trace arrows point to cells with predecessor or dependent references. All trace arrows are cleared when either the Circular Reference toolbar is closed or the Remove All Arrows button on the right side of the Circular Reference toolbar is clicked.

Understanding circular references can be very helpful to you. Using circular references can be very powerful and timesaving if they are made intentionally. If made accidentally, it can take hours to trace the errors and correct them. Using the trace arrows in the Circular Reference toolbar can save a lot of time and frustration when setting up your spreadsheets.

Summary

Loan interest and builder's margin are not often estimated as accurately as they should be, especially when considering the amount of profit that can be lost without realizing it. True, you can't raise prices just because you want a higher profit margin, but at least you should know, up front, what your real profit margin will be when you contract for a project at a specific price. The understanding that you have gained in this chapter will make it possible for you to accurately calculate interest and to correctly establish an appropriate builder's margin. The spreadsheet tools you have learned here will, when implemented, enable you to quickly analyze what-if scenarios and make decisions that will help you price jobs so that they are profitable.

Automating Spreadsheets with Macros

CHAPTER 8

IN THIS CHAPTER

Learn how to:

- Create macros.
- Execute (run) macros.
- Create hot keys to run macros.
- Attach macros to macro buttons.
- Attach macros to drawings, pictures, clip art, scanned images, or other objects.
- Create custom toolbars.
- Create custom Menu items.
- Edit macros in the VBA editor

The previous chapters dealt with functions and formulas that can be used to automate estimating. Using macros to automate can be a powerful way to increase the efficiency and effectiveness of spreadsheets. A macro is a program you write or record that stores a series of commands that you can later use as a single command. Macros can be used for simple or complex tasks. Whenever an Excel task becomes repetitive, a macro can be created to accomplish the same task with only the click of a button.

Macros can be used to simplify tasks such as:

- Entering business information.
- Custom formatting of cells.
- Data manipulation.
- Creating reports.
- Performing mathematical computations.
- Etc.—The applications for using macros are endless.

Excel's macro language is Visual Basic for Applications (VBA). It can take a lot of time and knowledge to write a macro using VBA programming language, but Excel has taken most of the work out of the process. Excel has a macro recorder that records each step as you perform a command or a series of commands. To create a macro, the macro recorder is turned on, the steps are performed exactly in the order that you want a task completed, and then the recorder is turned off. Once a macro has been recorded, it can be attached to an object such as a drawing, a picture, an icon, a toolbar, or a menu item, then whenever the object that the macro is attached to is mouse-clicked, the macro executes (runs).

Macros

Mike, from New Home Construction, wanted to create a macro that would automatically enter his company's name and address in any location on a spreadsheet that he chose. His company's name and address are:

New Home Construction
550 North Star Valley Drive
Dallas, Texas 75382-2877

He was able to create a macro that would enter his company's name and address with one click of the mouse button. The task was simpler than he thought. Follow along to see how easy it is to create a macro.

To begin recording New Home Construction address macro:

1. Click **Tools**.
2. Click **Macro**.
3. Click **Record New Macro**. The **Record Macro** dialog box pops up (Figure 8.1).

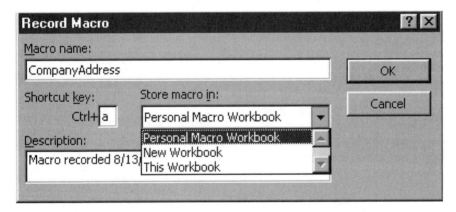

Figure 8.1 Recording a Macro

4. Enter the one-word name for the macro (CompanyAddress) in the Macro name box.
5. A shortcut key can be entered which allows the operator to run the macro just by pressing the control key at the same time as the shortcut key. To setup a shortcut key using the letter a for address, enter a next to the **Ctrl+** box under Shortcut key.

Once completed, the macro can be executed by pressing the letter a while holding down the **CTRL** button (**CTRL+a**).

The **Store macro in:** box asks where the macro should be kept. If you want the macro to be available whenever Excel is open, then choose **Personal Macro Workbook**. Store the **CompanyAddress** macro in the **Personal Macro Workbook** so that it will be available no matter which workbooks are open. For macros that are needed only when a specific workbook is open, choose one of the other workbook options, usually **This Workbook** option. If your computer spreadsheet will be used on another computer, then the macro must be stored on **This Workbook** or **New Workbook** and not the **Personal Macro Workbook**.

When all of the required information has been entered into the **Record** macro dialog box, click the **OK** button to start recording. **Recording** will display on the status bar at the bottom of the screen and the **Stop Recording** toolbar will pop up (Figure 8.2).

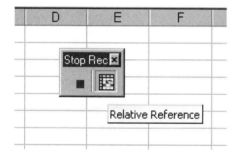

Figure 8.2 The Stop Recording Toolbar

The **Relative Reference** button, on the right side of the Stop Recording toolbar, allows you to record a macro using relative references or using absolute references. A relative reference (position depressed as in Figure 8.2) runs the macro in relation (relative) to the position of the cursor. We would want the ReturnAddress macro to begin at the position of the cursor, and so, would turn the relative reference on.

Click on the **Relative Reference** button.

While "Recording" is displayed in the status bar (near the bottom of the screen), every keystroke and click that you make in Excel will be recorded. Without moving the position of the cursor, begin entering the macro by typing the name and address of New Home Construction (Figure 8.3). It doesn't matter the position of the cursor when you begin

Figure 8.3 Creating a Macro to Enter the Company Name and Address

entering the macro; the important thing to remember is not to position the cursor after the recording has started because then the macro would always start where the cursor had been moved.

1. Type New Home Construction
2. Press **Enter**
3. Type 550 North Star Valley Drive
4. Press **Enter**
5. Type Dallas, Texas 75382-2877
6. Press **Enter**

When the company address has been entered the way you want it to look, click the **Stop Recording** button on the left side of the **Stop Recording** toolbar (Figure 8.3) and the **Stop Recording** toolbar will disappear.

> **TIP**
>
> Tip: If the Stop Recording toolbar has been closed, it will not display when a macro is in the record mode. To open the Stop Recording toolbar:
>
> Right click on any toolbar and click **Stop Recording** (Figure 8.4), or click **View/Toolbars** and click **Stop Recording** (Figure 8.4).

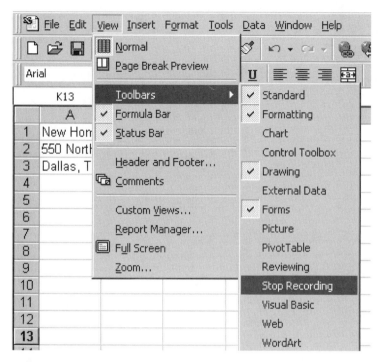

Figure 8.4 Opening the Stop Recording Toolbar

Running Macros

There are several ways to run or execute the **CompanyAddress** macro. One way is to:

1. Select the cell where you want to enter the company name and address (**B3** for example on a new worksheet).
2. Click **Tools**.
3. Click **Macro**.
4. Click **Macros** and the **Macro** dialog box appears. It contains the list of the macros in all of the open workbooks (Figure 8.5).

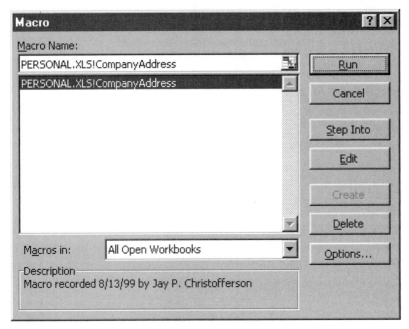

Figure 8.5 The Macro Dialog Box

5. Select **PERSONAL.XLS!CompanyAddress** from the list of open macros in the Macro dialog box.
6. Click the **Run** button, or simply double click on the macro name. The result should appear as the example in Figure 8.6.

Figure 8.6 Results from Running the CompanyAddress Macro from Cell B3

The **Personal.xls** workbook, shown in Figure 8.5, is a hidden workbook that automatically opens every time Excel is opened. If **Personal Macro Workbook** is selected when you create the macro (as was the case with **CompanyAddress** macro), then the macro will be stored in the **Personal.xls** workbook. Macros stored in the **Personal.xls** workbook can be executed any time Excel is open.

If the ReturnAddress macro had been recorded in Absolute Reference mode, it would have always run in the same cells, no matter where you put the cursor before you executed the macro. Try creating a unique address macro or label macro using the Absolute Reference mode and see how it functions differently.

The Shortcut Method of Running Macros

Another and quicker way to run the CompanyAddress macro is to use the shortcut key, **Ctrl+a**:

Select the cell where you want the company name and address to be located. While holding down the **Ctrl** key, press a.

Command Buttons

Macros can also be attached to any object so that when the object is clicked, the macro executes. One object made specifically for macros is the **Command** button icon on the **Forms** toolbar (Figure 8.7). To create a command button:

Figure 8.7 The Command Button Icon on the Forms Toolbar

1. Click the **Command** button (**Macro** button) on the **Forms** toolbar.
2. Move to the location on the worksheet where you want the command button to be positioned.
3. Click and drag to create and size the **Command** button on the sheet (hold the left mouse button down while moving to a new location).

The **Command** button appears on the worksheet and the **Assign Macro** dialog box appears (Figure 8.8).

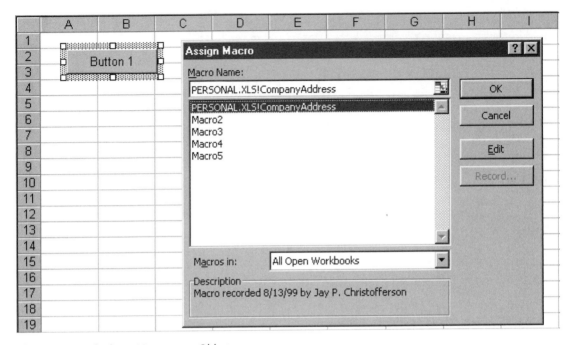

Figure 8.8 Assigning a Macro to an Object

1. Choose from the list of open macros the macro that you want to attach to the button object.
2. Click **OK**.

With the edit handles (small squares surrounding the macro button) showing, you can change the name of the button to Company Address or some other descriptive name by clicking on the text in the button.

Once you click off of the newly created macro button, it is activated (the edit handles disappear) and whenever you click on the command button, the CompanyAddress macro will be executed.

Editing Command Buttons

To again be able to edit the command button text, or move or resize the button, right-click on the command button. A shortcut menu pops up (Figure 8.9) which allows you to edit the text, edit the command button, or assign (reassign) a macro to the button.

Figure 8.9 The Shortcut Menu for Objects

Creating Your Own Command Buttons

Macros can also be assigned to pictures, clip art, scanned images, and drawing objects. To create a drawing object:

1. Open the **Drawing** toolbar (Click **View/Toolbar/Drawing** or right-click on any open toolbar and select **Drawing**)
2. You may choose from the drawing objects that are available on the **Drawing** toolbar (such as a square, oval, etc.). Click on the oval.
3. Click and drag on the worksheet in the location you where you want the object.
4. Right-click the object; the object's edit handles display as does the shortcut menu.
5. On the shortcut menu, select **Format AutoShape** to format the object (Figure 8.10).

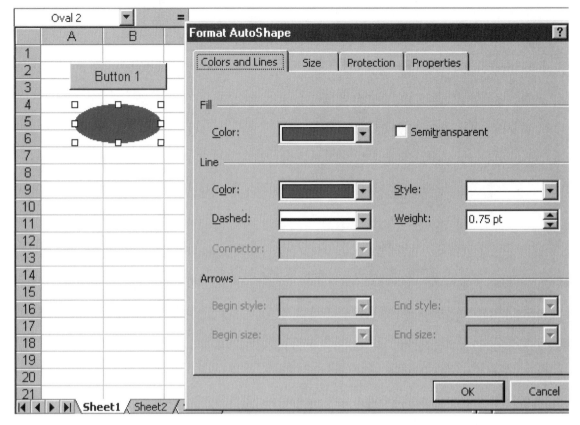

Figure 8.10 Changing the Appearance of an Object

6. Make the desired changes to the command button's appearance.
7. Click **OK**.

Attaching Macros to Objects

Once the object has been created, a macro has to be attached to it. To attach the CompanyAddress macro to the oval object:

1. Right click on the oval to open the shortcut menu
2. Choose **Assign Macro** (Figure 8.11).

Figure 8.11 Assigning a Macro to an Object

3. Select the **CompanyAddress** macro from the list of open macros.
4. When you click off of the oval, it becomes activated (the edit handles disappear). To test the macro:
5. Select the cell where you want the company's name and address to be entered.
6. Click the oval.

Creating Custom Icons

The toolbars that you have been working with in Excel are collections of icon objects that have macros attached. You can create your own custom icons that will execute your macros when you click on the icon buttons. To create your own custom toolbar:

1. Click **View**.
2. Click **Toolbar**.
3. Click **Customize**. The **Customize** dialog box appears.

4. Click **New**. The **New Toolbar** dialog box pops up and prompts you to give the toolbar a name (Figure 8.12).

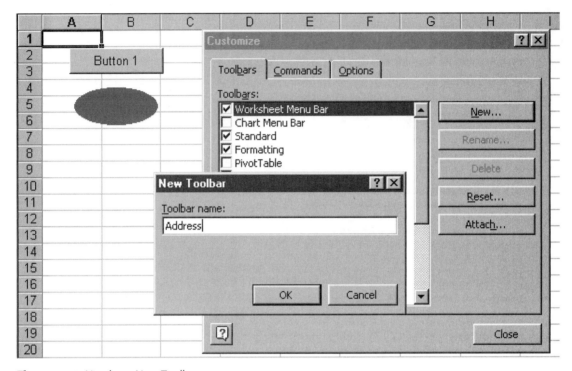

Figure 8.12 Naming a New Toolbar

5. Enter an appropriate name (Address).
6. Click **OK**.

The toolbar displays as a blank box (see the blank toolbar below the oval object). To add a button to the toolbar:

1. Select the **Commands** tab of the **Customize** dialog box
2. Select **Macros**.
3. Click the **Custom** button and drag it into the toolbar (Figure 8.13) and release the mouse button. The smiley face displays in the toolbar.

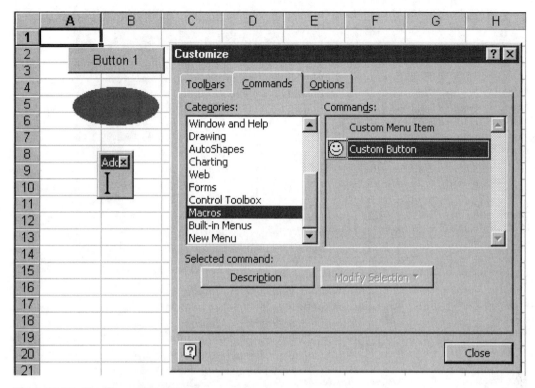

Figure 8.13 Creating a Custom Button

Changing Icon Images

If you like the smiley face, you can keep it as the CompanyAddress icon button, or you can choose another pre-designed button.

- With the **Customize** dialog box showing, right-click on the smiley face. A button menu displays.
- Click **Change Button Image** and an array of images appears (Figure 8.14). You may select the image you want or if none of them have the appearance you want, you can create a custom button image.

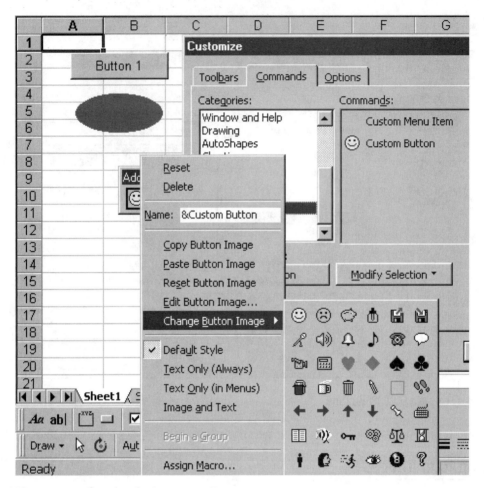

Figure 8.14 Changing the Image on a Button

If you want to create your own custom button, you can select a button image that resembles the image you want and then modify it, or you can select the empty image box and start from scratch to create your custom button image. For our custom button:

1. Select the empty box icon and it will display in the **Address** toolbar.
2. With the **Customize** dialog box still open, right-click on the empty box (inside the Address toolbar).
3. Click the **Edit Button** image on the menu. The **Button Editor** appears which enables bit-map editing of the image (Figure 8.15).

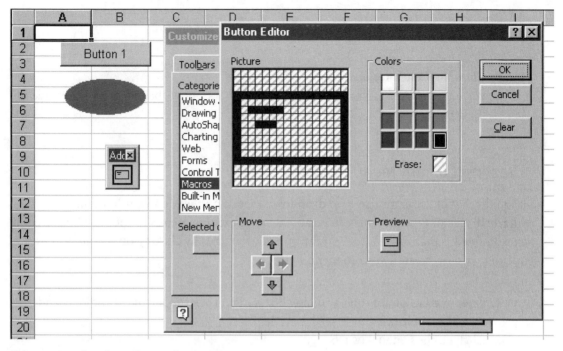

Figure 8.15 Creating a Custom Button Image

4. Draw an image.

You will want to create an icon image that represents the **CompanyAddress** macro. You may want to create the image of an envelope and return address as shown in Figure 8.15. Alternatively, you could design something creative on your own. Notice the preview screen near the bottom of the button editor. As you create the button image, you can see what the button will look like at its proper size.

5. When you have completed the image, click **OK**
6. Remember to attach the **CompanyAddress** macro to the icon image:

- With the Customize dialog box still open, right-click on the icon image in the **Address** toolbar.
- Click **Assign Macro**.
- From the list of opened macros, double-click on the name of the macro that you want to attach to the icon.

To finish the custom toolbar,

- Close the **Customize** dialog box.
- Place the **Address** toolbar with the other toolbars, above, below, or to either side of the screen.

Creating Custom Menu Items

Even the text of the items listed on dropdown menus on the Menu bar have macros attached. Custom menu items can be added to the existing menus or they can be placed in custom menus that you create. You can run your macros by clicking on one of the custom menu items. To create a custom menu item:

1. Open the **Customize** toolbar dialog box. (Right click on one of the toolbars and select **Customize**).
2. Click on the **Commands** tab.
3. Click **Macros** from the **Categories** box.
4. Select **Custom Menu Item** from the **Commands** box.
5. Drag **Custom Menu Item** to the desired location in one of the menus or next to the menu selections.

To create a menu item called Company Address just below Macro on the Tools menu, drag **Custom Menu Item** to the **Tools** menu (the menu list displays) and continue to drag down to just below **Macro** and release the mouse button (Figure 8.16).

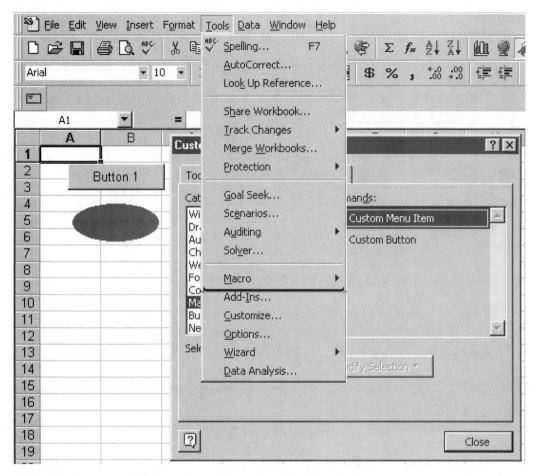

Figure 8.16 Positioning Bar to Insert a Custom Menu Item

Excel inserts **Custom Menu Item** in the place where you released the mouse button (Figure 8.17).

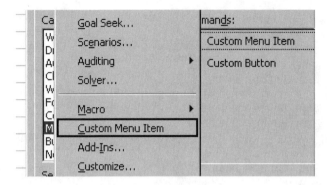

Figure 8.17 Custom Menu Item Inserted into the Tools Menu

If you accidentally drop the **Custom Menu Item** in the wrong place:

- With the **Customize** toolbar dialog box open, click and drag the menu item to a new location. If you want to delete the custom menu item, drag it back to the **Customize** toolbar dialog box.

It would be more descriptive to have a more appropriate name for the menu item than Custom Menu Item. To change the name to Company Address:

- With the Customize toolbar dialog box open, right click on **Custom Menu Item**. The **Shortcut** menu pops up (Figure 8.18). This

> **TIP**
>
> The menu items with the underlined letters in their names can be activated by pressing the **ALT** key at the same time as the underlined letter (**ALT+c** for **Custom Menu Item**).

Automating Spreadsheets with Macros

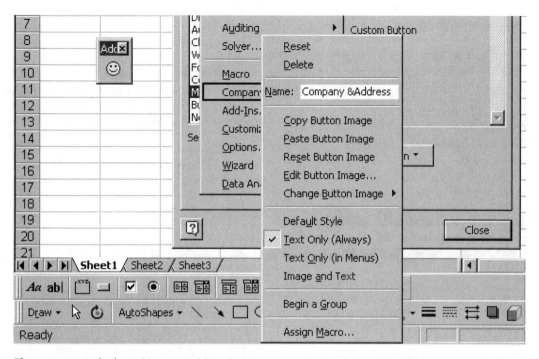

Figure 8.18 Assigning a Macro to a Menu Item

menu allows you to rename and assign a macro to the menu item.
- In the name box, type Company & Address. If you want to use a hotkey (**Alt+a**) to initiate the **CompanyAddress** macro, enter an ampersand (&) just in front of the hot key (A).

You will also need to attach the macro CompanyAddress to the Company Address menu item.

If the **Customize** toolbar dialog box is not already open, right-click on **Custom Menu Item**. The **Shortcut** menu pops up (Figure 8.18).

1. Select **Assign Macro** from the shortcut menu.
2. Double-click on **PERSONAL.XLS!CompanyAddress**.

Editing Macros in Visual Basic for Applications

The macro that we recorded (CompanyAddress) was written as a procedure in Visual Basic for Application's programming language. You really don't need to know how to write in VBA to be able to create and use macros, but it is handy sometimes to be able to edit the procedural code and make corrections or adjustments.

Because **Personal.xls** is a hidden file, we must unhide the file before we can see the code for **CompanyAddress**. To unhide the **Personal.xls** file:

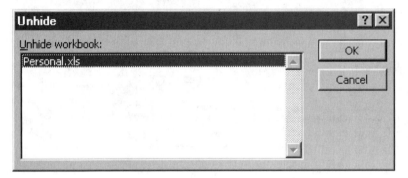

Figure 8.19 Making a Workbook Visible

1. Click **Window**.
2. Click **Unhide**. The Unhide dialog box appears (Figure 8.19).
3. Double-click on **Personal.xls** to unhide it. To check that **Personal.xls** is visible, click **Window** to view the open files.

The Visual Basic Editor opens automatically when any macro is edited. To open the Visual Basic Editor to the CompanyAddress procedure (macro):

1. Click **Tools**.
2. Click **Macro**.
3. Click **Macros**.
4. Select **CompanyAddress**.
5. Click **Edit**. The Visual Basic Editor opens to the **CompanyAddress** macro (Figure 8.20).

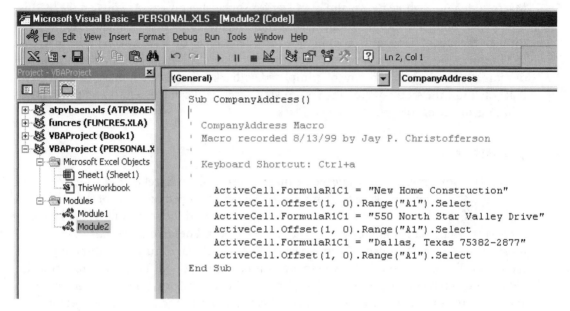

Figure 8.20 Editing Macros through the Visual Basic Programming Language Editor

6. Macros are stored in modules that are connected to workbooks and worksheets. Statements that are preceded by apostrophes ("'") are not code, but are comments.

To edit the text:

1. Move the cursor into the code.
2. Make the desired changes. For example, change Drive to Dr. and Texas to TX.
3. Do not press enter after you make the changes, but click on the Excel tab at the bottom of the screen.
4. Select the cell where you want the name and address of the company to be entered.
5. Run **CompanyAddress** using any one of the methods that you have learned earlier in the chapter and check to see that the macro is executing properly. When the macro works properly, save the **Personal.xls** file.

Macros open up a whole new realm of power for spreadsheet users. As you can see by this example, creating a macro is as simple as turning the macro recorder on, stepping through the task at hand, and stopping the recorder. Once you have attached the macro to a button, object, hotkey, or menu item, the task that may have taken several minutes, or more, to accomplish, can now be completed with a key-press or a mouse click. Now that you know the secret of creating your own macros, you can create fast and powerful solutions to any of your repetitive Excel tasks.

Summary

Visual Basic for Applications can execute many commands in Excel that are not even possible from the normal worksheets that we use. We have touched on some of the macro basics in this book; to delve deeper into macros goes beyond the scope of this book, but macros are a fascinating and powerful feature of Excel and deserve extra time to learn more about. To check out some very powerful and useful macros used in estimating, open the sample Estimator*PRO* on the CD that accompanies this book. The procedural code in VBA allows interaction between the program and the user by providing message, input, and dialog boxes. These allow users to select items from lists of choices and to perform operations based on those choices.

Using Form Tools to Enhance Your Spreadsheets

CHAPTER

9

IN THIS CHAPTER

Learn how to customize your spreadsheets using:

- Option Button
- Check Box
- Scroll Bar
- List Box
- Combo Box
- Spinner

Form tools are controls that provide additional power and utility when creating customized spreadsheets.

Form tools are controls that provide additional power and utility when creating customized spreadsheets. Option Buttons, Check Boxes, List Boxes and Combo Boxes let you quickly choose and calculate different combinations when estimating. Scroll Bars are an added feature that allow you to navigate more quickly and efficiently within your spreadsheet.

You can view the **Forms** toolbar by right clicking on any of the toolbars to show the **Toolbars** menu and then left click on **Forms**. The **Forms** toolbar will display on your screen and you can then move it where you want (Figure 9.1).

Figure 9.1 The Forms Toolbar

To move the toolbar:

1. Click on the (blue) name bar above the controls and (without releasing the click) drag the toolbar to the top, bottom, or one of the sides of your screen.
2. Release the click and the toolbar will stay.
3. If you have to adjust the position or move the toolbar again, click on the small bar near the first (left or top) of the toolbar and drag the toolbar to the new position.

Forms toolbars add power to your spreadsheets and some can be very useful. The **Option Button** and the **Check Box**, for example, can be especially helpful.

Option Buttons

Suppose, for example, that you wanted to select from the options of using a ¾-, 1-, or 1½-inch main water supply line. The connection fee for the ¾-inch line is $800.00, $1,750.00 for the 1-inch line, and $3,500.00 for the 1½-inch water line. You can use the Option Button form tool to accomplish this task. To set up the Option Button:

1. Click the **Option Button** on the Forms toolbar. The mouse pointer changes to a crosshair.
2. Click on the worksheet where you want the top left of the **Option Button** to be located and drag diagonally to the right and down to the desired size of the **Option Button** and release the click (Figure 9.2).

Using Form Tools to Enhance Your Spreadsheets **137**

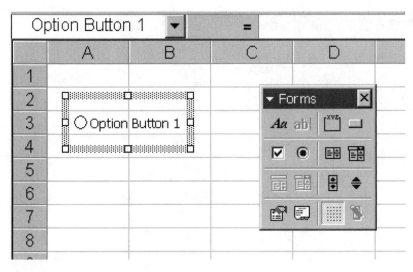

Figure 9.2 Creating an Option Button

To move or resize the **Option Button**:

1. Right click with the mouse pointer over the **Option Button**.
2. Press the **Escape** (Esc) button to clear the menu.
3. To move, click on the edge of the border and drag to the desired location
4. To resize, click on one of the box handles and drag to the desired size.
5. Repeat step 2 for the number of option buttons that you want to create, or copy the **Option Button** by placing the mouse pointer over it, right click, select **Copy**, right click over a cell near the location where the new **Option Button** is to be placed, and select paste.
    Alternatively, the copy operation can be accomplished by right clicking with the pointer over the **Option Button**, clearing the menu by pressing the **Escape** button, and while holding the **Ctrl** key down, click and drag the border of the option button to the desired location.

After the option buttons have been created, they need to be referenced to a cell on the worksheet.

1. Right click on any one of the option buttons and select **Format Control** from the pop-up menu (Figure 9.3).

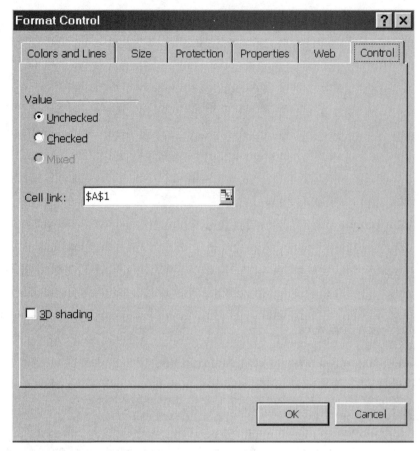

Figure 9.3 Formatting the Option Button Control

2. Click the **Control** tab and select the box next to **Cell link**.
3. If you wanted the cell reference to be located in cell **A1**, enter the cell address or click on the worksheet cell (**A1**) and the cell address will automatically be entered into the **Address Reference** box.
4. Click **OK**.

The value of cell **A1** will be blank if none of the option buttons have been clicked. If one of the option buttons is selected (the button is black), the value of the cell link (**A1**) will be 1, 2, or 3. Only one button at a time can be selected.

To label the option buttons:

1. Right click on each button and select **Edit Text** from the menu.
2. Change the text from **Option Button 1**, for example, to **¾-inch water line**.

Now, to make the option buttons work, we will use an IF-Then statement in a formula.

1. Enter the IF-Then formula into cell **C1**.
2. The formula could be stated, If **Option Button 1** is checked (¾-inch water line), then the water line cost is $800.00 or else if **Option Button 2** is checked (1-inch water line), then the water line cost is $1,750.00 or else use the only other option—the 1½-inch water line that costs $3,500.00.

The formula in cell **C1** would be written as follows (Figure 9.4):

=IF(A1=1,800,IF(A1=2,1750,3500))

	A	B	C	D	E
1	2	○ 3/4" Water Line	$ 1,750.00		
2		⦿ 1" Water Line			
3		○ 1-1/2" Water Line			
4					

Figure 9.4 Using an IF-Then Statement with Option Buttons

Cell **A1** stores the value of the option button that is selected (see number 4 above).

TIP

So that the text in cell A1 doesn't distract from the worksheet, it can be turned off (change the text color to white) so that it is no longer visible on the screen.

TIP

If you want two or more sets of option buttons on a worksheet, you will need to group each set so that each set will act independently of other option buttons on the worksheet. Group each set by selecting the **Group Box** button (the square-looking icon with XYZ written at the top) from the **Forms** toolbar then clicking and dragging on the worksheet so that a group box surrounds each set of option buttons Figure 9.5. The option buttons in each group will link to the same cell for that group and the option buttons for each other group will link to their own distinct cell. (See number 4 above.)

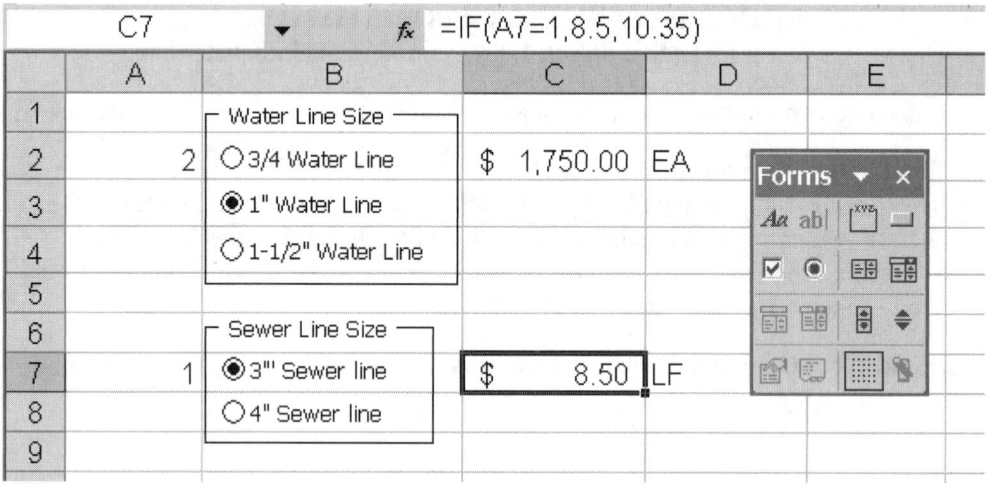

Figure 9.5 Using a Group Box to Cluster Option Buttons

Check Boxes

Check Boxes are similar to Option Buttons but any one or all of the check boxes on a worksheet may be checked. When the check box has a check mark in it, its value is True and when the check box is empty, the value is False. The value can be stored in a linked cell and each check box is linked to a separate cell (unlike option buttons where multiple option buttons are linked to the same cell). A good use for check boxes might be for answering yes or no questions about an item. For example, when buying a window you may want to select from the features that are available; you may want the option of ordering your windows with low E glass, with tinted glass, or with grids?

Look at the following example (Figure 9.6). Windows are listed with their base unit costs. Check boxes can be used to automatically calculate additional costs for low E glass, tinted glass and grids. The Unit Total is automatically calculated (Unit Cost + each of the additional costs). The Total Cost is also automatically calculated (QTY x Unit Total).

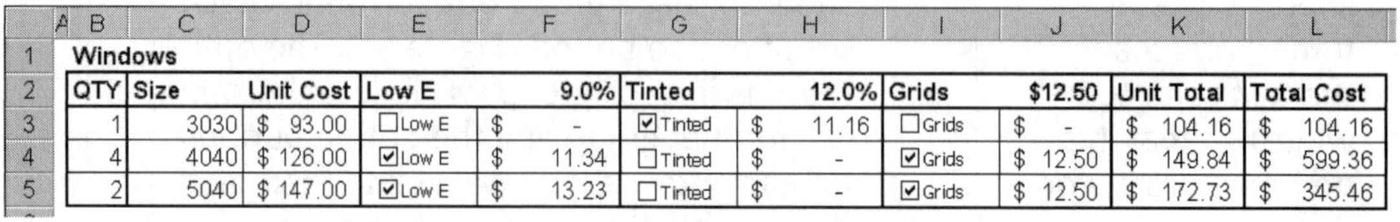

	A	B	C	D	E	F	G	H	I	J	K	L
1		Windows										
2		QTY	Size	Unit Cost	Low E		9.0%	Tinted	12.0%	Grids	$12.50	Unit Total	Total Cost
3		1	3030	$ 93.00	☐ Low E	$ -		☑ Tinted	11.16	☐ Grids	$ -	$ 104.16	$ 104.16
4		4	4040	$ 126.00	☑ Low E	$ 11.34		☐ Tinted	$ -	☑ Grids	$ 12.50	$ 149.84	$ 599.36
5		2	5040	$ 147.00	☑ Low E	$ 13.23		☐ Tinted	$ -	☑ Grids	$ 12.50	$ 172.73	$ 345.46

Figure 9.6 Using Check Boxes

Use the example in Figure 9.6 to begin setting up the check boxes.

1. Click the **Check Box** in the **Forms** toolbar and click and drag to the desired size (start by creating the **Check Box** in cell **E3**).
2. Change the name of the label to Low E by right clicking on the **Check Box** and clicking **Edit Text**.
3. Reposition the **Check Box** where you want by right clicking on the **Check Box**, press the **Escape** button on your keyboard, and then press the appropriate arrow keys on your keyboard to move the **Check Box** (or click on the border of the **Check Box** frame and drag to the new location).
4. Once the first **Check Box** is created and labeled correctly, you can copy it to create the check boxes below it.

The next step is to create a cell link for each check box. Remember that the value of the check box (True if checked and False if left blank) can be linked to a cell in the worksheet. To create the link:

1. Right click on the **Check Box**, and click **Format** Control from the menu.
2. Click the **Control** tab (on the Format Control pop-up box) and click in the **Cell Link** edit box. The cell address can be typed directly into the edit box. For an alternative and perhaps quicker method to enter the cell address, click on the button to the right of the edit box (it reduces in size) and click the worksheet cell that you want to link to, and then press enter.
3. Click **OK** to finalize the cell link.

Look at the Check Box over cell E3, for example (Figure 9.6). A good cell to link to would be E3 (the linked cell just behind the Check Box). When the box is checked, the value of E3 is True and False when left unchecked. To hide the display in E3 so that the displayed value doesn't overlap and clutter the checkbox, change the text color to white.

In F3, enter a formula that will use the linked information stored in E3. If adding Low E coating to the window order increases the cost of the base unit by nine percent (value stored in cell F2), the formula to correctly calculate the amount could be entered into cell F3:

=IF(E3,D3*F2,0)

Stated verbally, the formula would read: If cell E3 is True IF(E3, … is the same as if written =if(E3 = True, …), then multiply the base unit cost (D3) by .09 (F2) or else (if the value is False) leave the value as 0.

> **NOTE**
>
> The formula could also be written, =IF(E3,D3*$F$2,0). The $ sign in front of the F and the 2 anchors the reference so that as the formula is copied down to other cells, the reference to F2 does not change as do the other relative references (Figure 9.7). Remember, a quick way to anchor a cell reference is to place the cursor in or next to the reference and press the function key, F4. The function key, F4, is a toggle key and as you continue to press F4, the reference will change from F2 to $F$2 to F$2 to $F2 and then start the cycle again at F2.

F
0.09
=IF(E3=TRUE,D3*$F$2,0)
=IF(E4=TRUE,D4*$F$2,0)
=IF(E5=TRUE,D5*$F$2,0)

Figure 9.7 Anchored References in Formulas—$F$2

Scroll Bar

Scroll Bars can be helpful for inputting numbers or quantities into a cell. They are fun to create and can spice up a worksheet by adding functionality. They work similarly to the scroll bars on the side and bottom of MS Windows programs. By clicking on the arrow, the value of the linked cell changes by one (incremental change). By clicking inside the arrows, on the bar, the change is greater (page change). You can also drag the scroll bar to the desired value.

To create a scroll bar on your worksheet:

1. Click the **Scroll Bar** icon on the Forms toolbar and click and drag in the area over the worksheet where you want to place the **Scroll Bar**. The **Scroll Bar** must be linked to a cell for the value of the **Scroll Bar** to display on the worksheet.
2. Right click the **Scroll Bar** and the **Format Control** screen appears (Figure 9.8). Be sure the **Control** tab is selected.

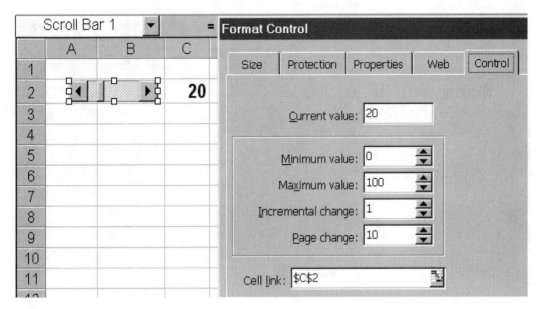

Figure 9.8 Using Scroll Bars

Enter the cell link address by clicking on the button to the right of the **Cell Link Edit** box and clicking on the desired cell (**C2** in this case). While the **Format Control** screen is still visible, set the minimum, maximum, incremental change and page change values to best fit your application.

Spinner

The **Spinner** works in a similar fashion to the Scroll Bar except that its value only changes incrementally. Suppose, for example, that you wanted to use a spinner to select the slope that would be used to calculate the squares of roof shingles (Figure 9.9).

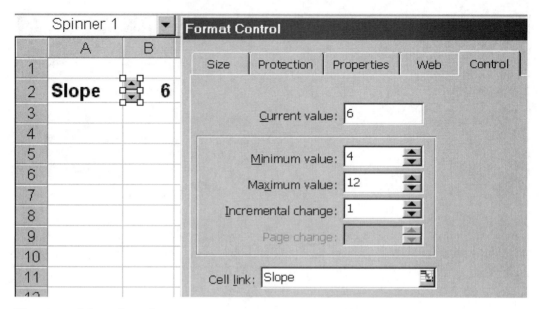

Figure 9.9 Spinner Control

Before making the **spinner control**, name cell **B2** Slope (click on **B2** and type Slope in the **Name** box and press **Enter**). To create the **spinner**:

1. Click the **spinner** icon on the **Forms** toolbar and then click and drag over the area on the worksheet where you want the **spinner** to be located.
2. Right click the **spinner** to format the controls for the **spinner**. You must set the minimum value (minimum slope you want to allow), the maximum value (the maximum slope you want to allow), the incremental change, and the cell link (type the named range Slope in the **Cell Link** box.
3. Click **OK** at the bottom of the **Format Control** pop-up screen). Cell **B2** (Slope) now displays the value of the spinner.

List Boxes

List Boxes are used to select items from a list. A List Box (Figure 9.10) displays all of the items assigned to that list.

Using Form Tools to Enhance Your Spreadsheets 145

![List Box showing dishwasher models in cells A3-A8: WP DU9700XR, WP DU9400XT, WP DU9200XT (highlighted), WP DU9000XR, GE Model GSD 1000T, GE Model GSD900D, GE Model GSD600D, GE Model GSD500D]

Figure 9.10 List Box

A List Box control can be formatted so that when the user clicks on an item in the list, the value of the position of that item is returned to a cell that has been previously designated. For example, suppose you had a list of dishwashers and had named the range of cells that contained the list of dishwashers, Dishwashers. You would create a List Box by:

1. Selecting the **List Box** button from the toolbar and clicking and dragging on the worksheet where you want the **List Box** to be located.
2. Then, right click on the **List Box** to bring up the **Format Control** pop up box (Figure 9.11).

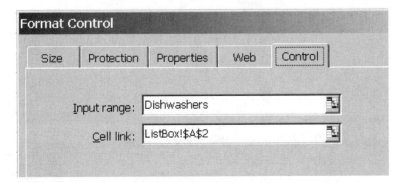

Figure 9.11 Format Control for List Box or Combo Box

3. Enter the named range, Dishwashers, into the input range, and enter the cell link (**A2**, in this case) into the **Cell Link** input box (Figure 9.12). The cell link is the location where the value of the selected item will be stored (cell **A2**).

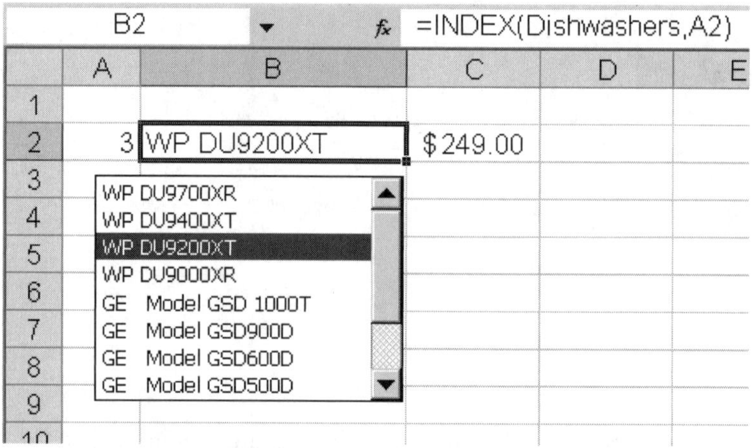

[Figure 9.12 shows a spreadsheet with cell B2 selected, formula bar showing =INDEX(Dishwashers,A2). Cell A2 contains 3, B2 shows WP DU9200XT, C2 shows $249.00. A dropdown list displays: WP DU9700XR, WP DU9400XT, WP DU9200XT (highlighted), WP DU9000XR, GE Model GSD 1000T, GE Model GSD900D, GE Model GSD600D, GE Model GSD500D.]

Figure 9.12 Making a List Box Work

> **NOTE**
>
> You can type in **A2** into the **Cell Link** input box or just click on cell A2 and the sheet name plus $A$2 will be automatically entered.

Knowing the position of the selected item is good but not very informative. To convert the position number to the name of the item (third item down), use the index formula.

= INDEX(Dishwashers, A2)

> **NOTE**
>
> The text in A2 can be turned off by changing the color to white.

Now that the item description is displayed in cell B2, you can make use of a VLOOKUP formula (in cell C2) to look up the cost from the database, DishwasherDB (review chapter 4). The formula that would be entered in cell C2 is:

=VLOOKUP(B2,DishwasherDB,2,FALSE)

If you wanted to, you could also look up the item's unit of measure (i.e. LF, SF, etc.) from the database using the VLOOKUP formula.

Combo Boxes

A **Combo Box** is similar to a List Box (Figure 9.13) and the same procedures for formatting the Input range and the Cell link apply, but when the focus is not on the Combo Box, it collapses so that only the selected item displays (Figure 9.14).

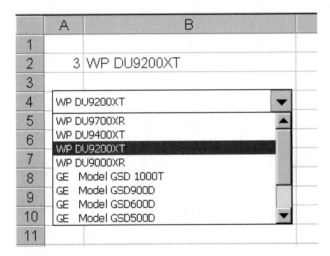

Figure 9.13 Selecting Items from a Combo Box List

Figure 9.14 After an Item is Selected from a Combo Box

The **Command** button is also a very useful Form tool and is primarily used for launching macros. You will learn more about Command buttons in the next chapter.

UserForms to Enter Data

CHAPTER

10

IN THIS CHAPTER

- Create UserForms to simplify data entry.
- Add your own controls to the UserForm.
- Quickly enter items for a roofing take off.
- Attach codes to the List Box and to other command buttons.

UserForms are powerful tools that increase the usefulness of Excel. UserForms are used as pop-up boxes that can display information, allow optional tasks, get information from the user, etc. **UserForms** are objects that can contain other objects (**Command Buttons**, **List Boxes**, **Labels**, **Spinners**, etc.), and each object can have code attached, called procedures, that tell the objects what to do. The code can get complicated but can also be very simple. It's worth taking a look into UserForms and getting a glimpse of the tremendous power behind Excel.

Data Validation

We will look at a very simple but powerful example that you can use with all of your detail sheets. In Chapter 4, you learned about Data Validation, which allows users to select items from a list. This can save a lot of time because the user doesn't have to type the name of the item, but instead, as the item is selected from a list, the name is automatically entered into a worksheet cell. Data Validation has some limitations when it comes to picking information from a list. One limitation is that for each cell that you want to enter information, you have to select the cell, open the list, and then you have to select the item. It would be faster and easier to open the list to select from and then be able to choose items while the list remains open and have the cursor automatically move down a cell after each selection.

Example

Suppose you had a roofing detail sheet that you wanted to be able to quickly enter the items that you would be taking off (Figure 10.1). You may already have the form set up to automatically return the Unit and Unit Cost for each item through VLOOKUP formulas (see chapter 4). You may also have entered a formula for calculating the Subtotal by multiplying the QTY by the Unit Cost.

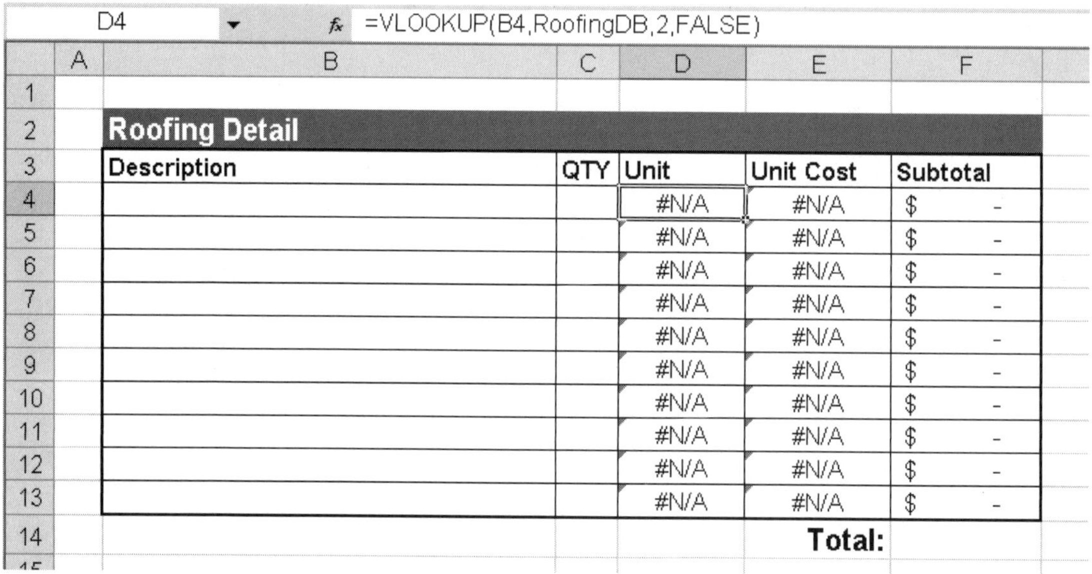

D4		fx =VLOOKUP(B4,RoofingDB,2,FALSE)			
A	B	C	D	E	F
1					
2	**Roofing Detail**				
3	Description	QTY	Unit	Unit Cost	Subtotal
4			#N/A	#N/A	$ -
5			#N/A	#N/A	$ -
6			#N/A	#N/A	$ -
7			#N/A	#N/A	$ -
8			#N/A	#N/A	$ -
9			#N/A	#N/A	$ -
10			#N/A	#N/A	$ -
11			#N/A	#N/A	$ -
12			#N/A	#N/A	$ -
13			#N/A	#N/A	$ -
14				Total:	

Figure 10.1 Detail Sheet with VLOOKUP Formulas

So that you know where we are going with this example, look at Figure 10.2. We want to be able to click the Select Material Command button and have the pop-up box (**UserForm1**) display the roofing database so that we can choose the specific items that we will use for the take-off on a project. By clicking the OK button once an item is selected and entered into the next cell on the worksheet. We also want to be able to select an item and enter it into the worksheet by double clicking on the item in the list.

UserForms to Enter Data **151**

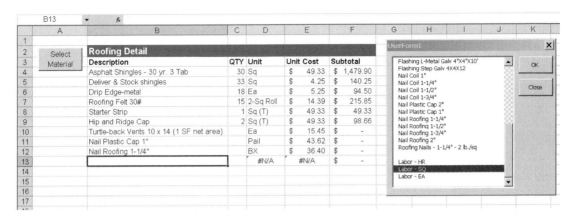

Figure 10.2 UserForms Can Be Used to Enter Item Descriptions

To begin this detail sheet, you will need to add headings, VLOOKUP and Subtotal formulas to the worksheet (Figure 10.3, also, review Chapter 4).

	B	C	D	E	F
1					
2	**Roofing Detail**				
3	Description	QTY	Unit	Unit Cost	Subtotal
4			=VLOOKUP(B4,RoofingDB,2,FALSE)	=VLOOKUP(B4,RoofingDB,3,FALSE)	=IF(B4="",0,C4*E4)
5			=VLOOKUP(B5,RoofingDB,2,FALSE)	=VLOOKUP(B5,RoofingDB,3,FALSE)	=IF(B5="",0,C5*E5)
6			=VLOOKUP(B6,RoofingDB,2,FALSE)	=VLOOKUP(B6,RoofingDB,3,FALSE)	=IF(B6="",0,C6*E6)
7			=VLOOKUP(B7,RoofingDB,2,FALSE)	=VLOOKUP(B7,RoofingDB,3,FALSE)	=IF(B7="",0,C7*E7)
8			=VLOOKUP(B8,RoofingDB,2,FALSE)	=VLOOKUP(B8,RoofingDB,3,FALSE)	=IF(B8="",0,C8*E8)
9			=VLOOKUP(B9,RoofingDB,2,FALSE)	=VLOOKUP(B9,RoofingDB,3,FALSE)	=IF(B9="",0,C9*E9)
10			=VLOOKUP(B10,RoofingDB,2,FALSE)	=VLOOKUP(B10,RoofingDB,3,FALSE)	=IF(B10="",0,C10*E10)
11			=VLOOKUP(B11,RoofingDB,2,FALSE)	=VLOOKUP(B11,RoofingDB,3,FALSE)	=IF(B11="",0,C11*E11)
12			=VLOOKUP(B12,RoofingDB,2,FALSE)	=VLOOKUP(B12,RoofingDB,3,FALSE)	=IF(B12="",0,C12*E12)
13			=VLOOKUP(B13,RoofingDB,2,FALSE)	=VLOOKUP(B13,RoofingDB,3,FALSE)	=IF(B13="",0,C13*E13)
14					

Figure 10.3 Automating Detail Sheets

UserForm

The next step is to add a button that can be used to open the UserForm. Any button that can be created can be used to execute a macro (see chapter 8). In this case, let's use the Command button from the Forms toolbar.

1. When you select the **Command** button on the **Forms** toolbar and click and drag on the worksheet to create a button, the **Assign Macro** pop-up box is displayed.

2. Click **New** to create the code that is needed to open the **UserForm**. The Microsoft Visual Basic Editor will open (Figure 10.4).

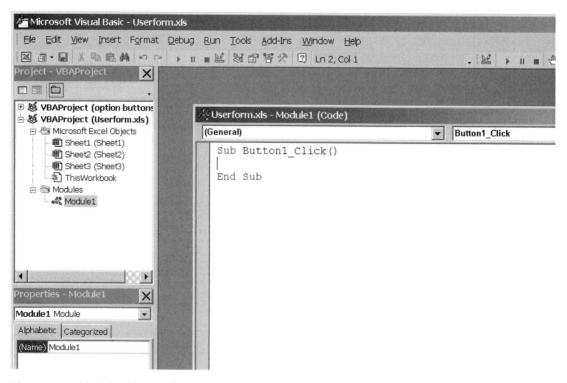

Figure 10.4 The Visual Basic Editor

You need to do two things at this point. You need to create a UserForm and also the code that is needed to open the UserForm from the worksheet. To create a UserForm:

1. Click **Insert** on the **Menu** bar.
2. Click **UserForm** (Figure 10.5). **UserForm1** is displayed together with a toolbox that contains the controls that are needed for adding buttons, list boxes, etc. to the **UserForm**.

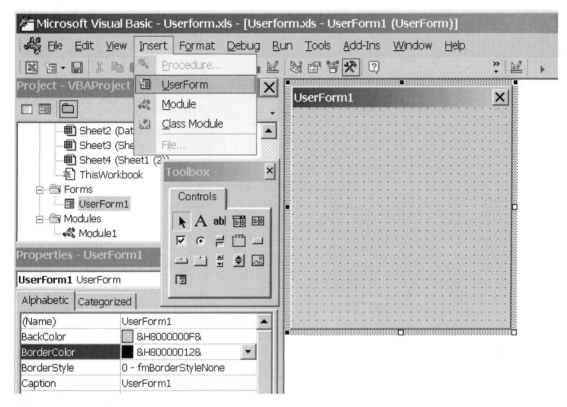

Figure 10.5 Creating a UserForm

Notice in the Properties box the name of the UserForm is **UserForm1** and the caption for the UserForm is also **UserForm1**. We could change either or both of these to be more descriptive, for example, the caption could be changed to **Material Database** or **Roofing Database**. We will leave them as is for this example.

To enter the code that is needed to display the UserForm:

1. Double click on **Module1** in the **Project** window.

Notice that the first line of the subroutine (the lines of code for a macro) has been entered and that the ending code is already in place. The code says that when **Button1** is clicked, that the lines of code that follow will be executed and when they are done, the subroutine will end.

2. Enter the code UserForm1.show (Figure 10.6).

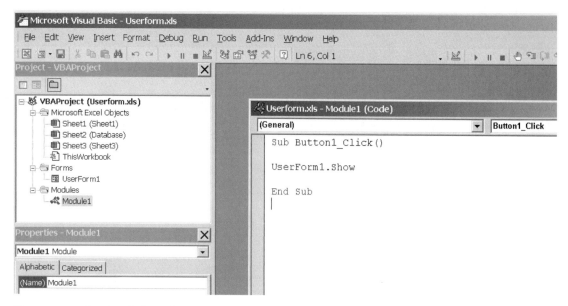

Figure 10.6 Creating Code to Open a UserForm

3. Press **Enter**.

Notice that when you enter the code and press the Enter key, some of the lowercase letters (u, f, and s) are automatically capitalized. This indicates that the VB editor recognizes the code that you have just typed and also the existence of the UserForm1. The code that you entered (UserForm1.show) tells Visual Basic to show UserForm1. When the button on the spreadsheet is clicked, **UserForm1** will be displayed.

List Boxes

You will need to add some controls to the UserForm to make it useful. Double click on **UserForm1** (in the **Project** window) to bring up the **UserForm**.

For this example, we are going to add a List Box and two buttons to the UserForm. A List Box will be used to display and select database items for the take-off. To enter a List Box

1. Click on the **List Box** tool in the **Controls** toolbox.
2. Click in the left, top of the **UserForm** and drag to the desired size of the **ListBox** (Figure 10.7).

UserForms to Enter Data **155**

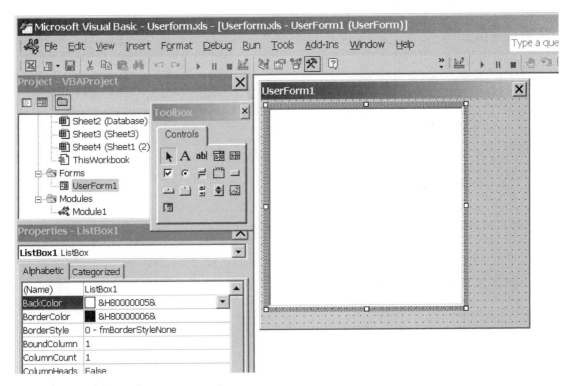

Figure 10.7 Adding a List Box to a UserForm

Two command buttons will be used on the UserForm, one to accept and process a selection from the List Box (the OK button), and one to close the UserForm. Add the two command buttons to the **UserForm** by:

1. Clicking the **Command** button on the **Controls** toolbox and
2. Click on the desired location on the **UserForm** and dragging to the chosen size (Figure 10.8). After the first button has been created, you can repeat this procedure to create the second button or you can copy the first button by holding the **Ctrl** key while clicking and dragging the first button (click on the edge of the button) to the new position. The new button is created when you release the mouse click.

156   Estimating with Microsoft Excel

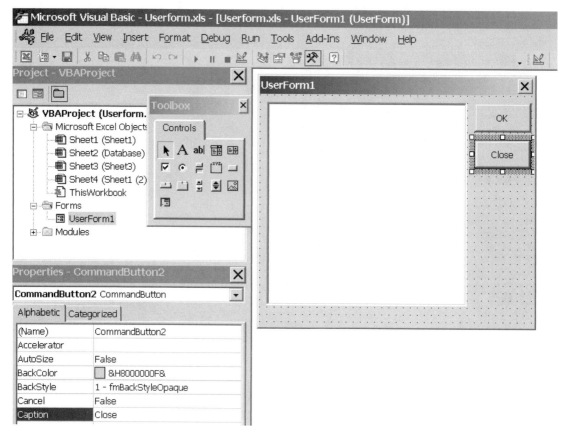

Figure 10.8 Adding Command Buttons to a UserForm

3. Change the button captions by changing the **Caption** text in the **Properties** window.

RowSource

You need to tell the List Box where to get the information it will use for its display. The **RowSource** property stores the link information for the List Box. A range of cells containing the names of item that you want displayed could be entered as the **RowSource**. A named range works even better as the RowSource.

To use a named range for roofing items, name the range of cells that contain the roofing item descriptions in the roofing database, **RoofingList**. Be sure that you first click on (select) the List Box and then, if you enter RoofingList as the **RowSource** property in the **Properties** window, the roofing list will fill the **List Box** with the items listed in the roofing database (Figure 10.9).

UserForms to Enter Data **157**

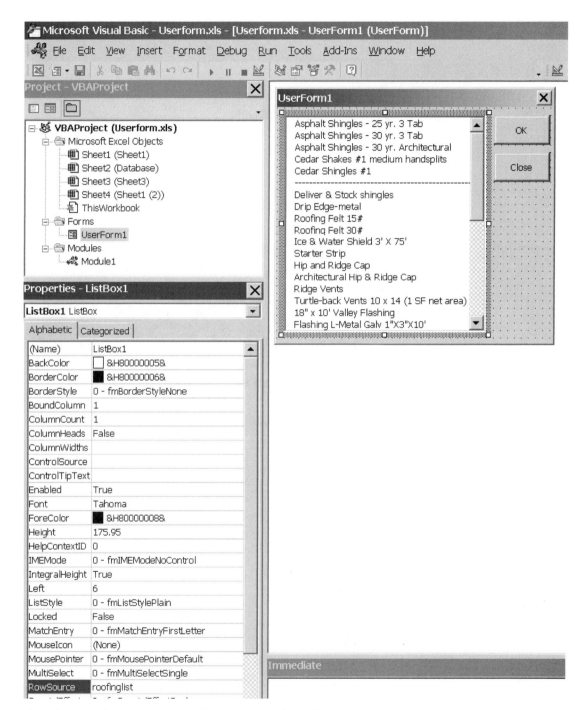

Figure 10.9 Using RowSource to Fill the ListBox with Roofing Items

158 Estimating with Microsoft Excel

You are now ready to attach code to the List Box and to the two command buttons. You will need to first, create the code that will take the information that is selected from the UserForm and enter it into the worksheet.

1. Double-click on the **OK** button and Module1 will display. This is the sheet that stores the code for the objects on the **UserForm**.
2. Type the following two lines just under **Private** Sub CommandButton1_Click() (Figure 10.10):

   ActiveCell.Value = ListBox1.Value
   ActiveCell.Offset(1, 0).Select

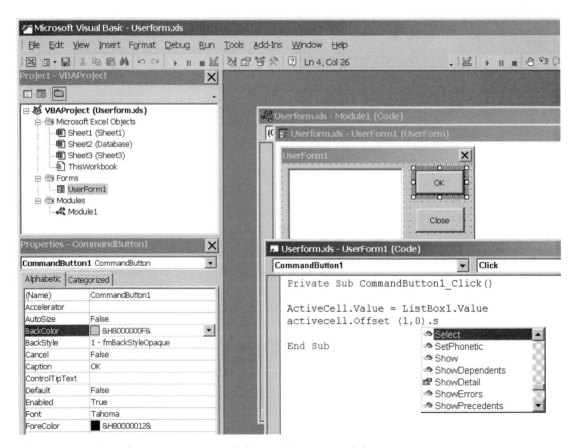

Figure 10.10 VB Code to Copy Item Description on ListBox to Worksheet

The first line of code tells Excel to take the value from the List Box and put it into the active cell on the worksheet. The active cell is the cell that is selected on the worksheet. The value is whatever has been selected on the List Box (not the position of the selected item in this case, but the actual description of the selected item in the list). The second line of code moves the active cell down one row. The first value in the parenthesis is the row offset value (1, in this case, meaning move down one row) and the second value in the parenthesis is the column offset value (0, because we are not changing columns).

> **NOTE**
>
> The command **ActiveCell.Offset (-1,-2).Select** would move the cursor up one row and two columns to the left.

Notice also, the help box next to the code as you type the period separating the property (Offset) from the event (Select). The help box also appears after the period is typed between an object identifier (such as ActiveCell) and a property.

To create the code that will close the UserForm when the close button is clicked, double click the **Close** button (Figure 10.11). Type the name of the object (UserForm1) followed by a period and then the method (Hide).

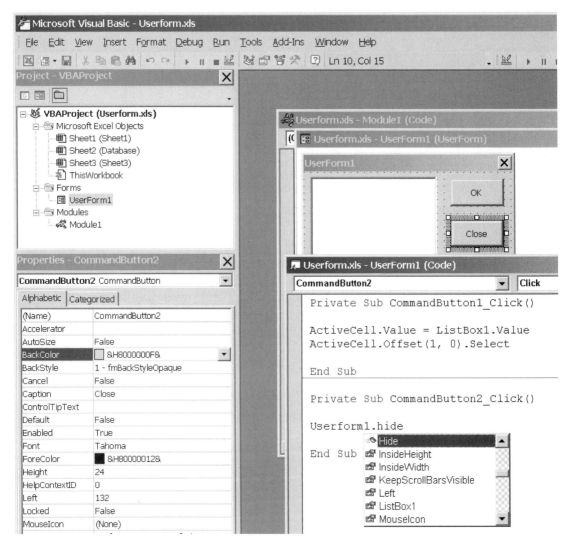

Figure 10.11 Closing a UserForm

To complete the code for our UserForm, you need to do one last thing. To be able to double click an item in the List Box and have the item's description appear in Excel in the selected cell, do the following; first:

1. Double click on the **List Box** to open the **UserForm** code editor. VBA automatically creates the beginning and ending code enabling a click event, but we want to double click an item before the code is executed (Figure 10.12).

UserForms to Enter Data   **161**

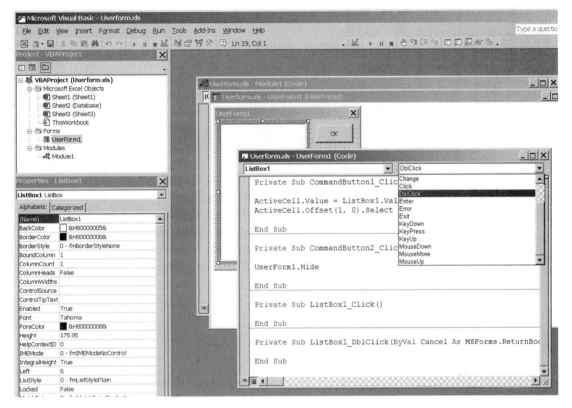

Figure 10.12 Creating a Double Click Event

To cause the code to be executed with a double click instead of only a single click, we need to change the event from **Click** to **DblClick** (top right of **UserForm1** code).

2. Click the down arrow that lists the events to display the possible events that can be used to execute the code.
3. Click on **DblClick** in the list of events. New lines of instruction are created that will run your code when you double click the **List Box**.

Because you want to run your code by double clicking an item in the List Box and not by single clicking, you can delete the ListBox1_Click() procedural code.

Now you can copy the same code that you used for CommandButton1 to the ListBox1_DblClick procedure (Figure 10.13).

162 Estimating with Microsoft Excel

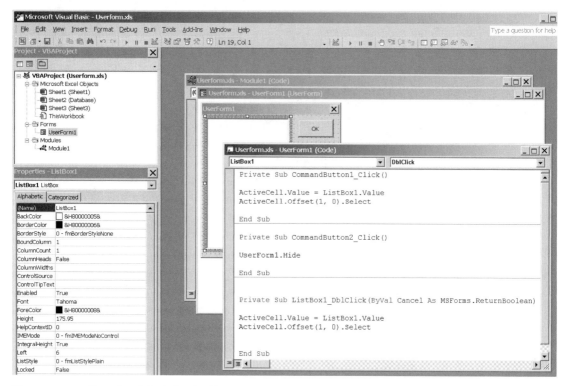

Figure 10.13 Completed Code for the UserForm

Test it out now and see how it works for you. Don't be afraid to try different settings and add your own controls to the UserForm or your own code to the procedure modules. Hopefully, this chapter has given you a sample of the remarkable power of Visual Basic for Applications that comes with MS Excel.

Summary

By now you have learned many of the basics and not so basics of Microsoft Excel. You have learned powerful functions and how to create practical formulas. You have learned how to create customized spreadsheets that will not only automate your estimating, but will reduce the time that you spend on many office tasks. The time you spend now to implement these time-saving techniques into your business will be as important as the maintenance that you perform on your equipment and the time you spend sharpening your saw. In this competitive environment, the tools that help you to become more effective and efficient will also make you more profitable.

One of the biggest complaints of working with computers is that information from one program has to be hand-typed into the next program for further processing. People are always asking for a program that does everything because they don't want to re-enter data between software programs. What they don't know is that most programs already have the ability to integrate. In the next chapter, we will look at how you can integrate with other types of software to increase computer effectiveness.

Integration: Using Excel with Other Programs

CHAPTER 11

IN THIS CHAPTER

Find out how integrating programs with Excel can:

- Increase productivity.
- Save data input time.
- Acquire data from other software programs.
- Process data internally.
- Send the information to other applications.
- Help your business run smoother.

Just like any useful tool, Excel can do many things, but it is not designed to do everything. On its own, Excel is powerful, versatile, and easy to use. A hammer is powerful, versatile, and easy to use too, but you wouldn't want to build a whole house using it as your only tool. You may already have many software tools for building your business or you may be in the process of obtaining key software tools to help your business run smoother. Using software tools independently can be helpful and will save time, however, using these tools in combination with other software programs can dramatically increase productivity and save data input time. This capability for Excel to integrate with other programs makes it possible to acquire data from other software programs, process this data internally, and send the information (processed data) on to other applications.

Other Applications

Many software programs can save data in the form of ASCII or text files. Figure 11.1 is an example of a database for house plans with information pertaining to plan identification, plan name, plan style, finished square feet, total square feet, base price, and the name of the architect. Each row is a record that stores information about a specific plan. Each record will contain different fields of information such as name, address, phone, size, etc. In this text file, each field is separated by a comma, which means that the file is comma-delimited.

```
Plan ID,Plan Name,Plan Style,FSF,TSF, Base Price, Architect
R-1400,Abilene,Rambler,1400,2600, 105000 ,Hargreaves
R-1825,Riverton,Rambler,1825,3500, 136875 ,Technigraphics
R-1980,Ranch ,Rambler,1980,3740, 148500 ,Hargreaves
R-2320,Windsor,Rambler,2320,4400, 174000 ,Technigraphics
SE-1050,Lancaster,Split Entry,1050,1600, 68250 ,Technigraphics
SE-1230,Cambridge,Split Entry,1230,1960, 79950 ,Design Specialists
SE-1500,Dearborne,Split Entry,1500,2500, 97500 ,Hargreaves
SL-1300,Sumpter,Split Level,1300,2100, 91000 ,Design Specialists
SL-1454,Olympic,Split Level,1454,2408, 101780 ,Technigraphics
SL-1672,Lexington,Split Level,1672,2844, 117040 ,Hargreaves
SL-1896,Uintah,Split Level,1896,3292, 132720 ,Technigraphics
TS-1850,Vermont,Two-Story,1850,2850, 138750 ,Technigraphics
TS-1930,Manchester,Two-Story,1930,2900, 144750 ,Design Specialists
TS-2120,Newberry,Two-Story,2120,3250, 159000 ,Technigraphics
TS-2340,Charlotte,Two-Story,2340,3320, 175500 ,Hargreaves
TS-2560,Richmond,Two-Story,2560,3580, 192000 ,Technigraphics
```

Figure 11.1 Comma-Delimited Text File

The fields that make up each record can also be separated by tabs as in Figure 11.2. When fields of a record are separated by tabs, the file is said to be tab-delimited.

Plan ID	Plan Name	Plan Style	FSF	TSF	Base Price	Architect
R-1400	Abilene	Rambler	1400	2600	105000	Hargreaves
R-1825	Riverton	Rambler	1825	3500	136875	Technigraphics
R-1980	Ranch	Rambler	1980	3740	148500	Hargreaves
R-2320	Windsor	Rambler	2320	4400	174000	Technigraphics
SE-1050	Lancaster	Split Entry	1050	1600	68250	Technigraphics
SE-1230	Cambridge	Split Entry	1230	1960	79950	Design Specialists
SE-1500	Dearborne	Split Entry	1500	2500	97500	Hargreaves
SL-1300	Sumpter	Split Level	1300	2100	91000	Design Specialists
SL-1454	Olympic	Split Level	1454	2408	101780	Technigraphics
SL-1672	Lexington	Split Level	1672	2844	117040	Hargreaves
SL-1896	Uintah	Split Level	1896	3292	132720	Technigraphics
TS-1850	Vermont	Two-Story	1850	2850	138750	Technigraphics
TS-1930	Manchester	Two-Story	1930	2900	144750	Design Specialists
TS-2120	Newberry	Two-Story	2120	3250	159000	Technigraphics
TS-2340	Charlotte	Two-Story	2340	3320	175500	Hargreaves
TS-2560	Richmond	Two-Story	2560	3580	192000	Technigraphics

Figure 11.2 Tab-Delimited Text File

To import a text file into Excel:

1. Click **File**.
2. Click **Open**.
3. Double click the file you want to import. When you open a text file (for example, a file with .txt, .csv, or .prn extensions), Excel's **Text Import Wizard** displays.
4. There are three steps to completing the import.

   - The first step is to indicate whether the fields are fixed widths, or delimited (separated) by commas, tabs, etc. (Figure 11.3). As you make decisions, the **Data Preview** box shows the results of those choices. When completed with this step click **Next**.

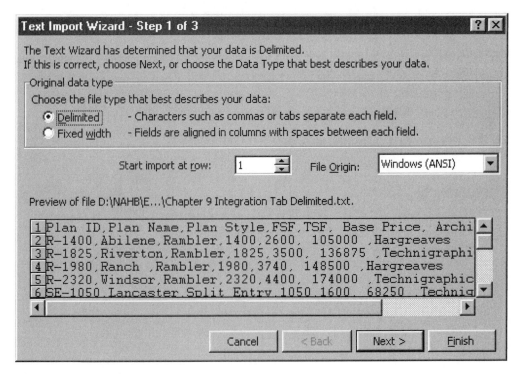

Figure 11.3 Importing Data, Step 1

   - Step 2 asks you to identify the type of delimiter used to separate the fields (Figure 11.4). Indicate if the fields are separated by tabs, commas, semicolons, spaces, or some other character. Click **Next** to move to the next step.

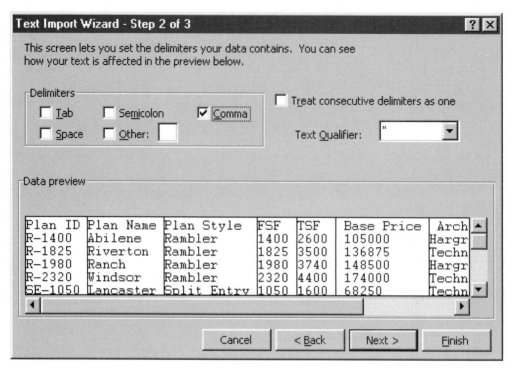

Figure 11.4 Importing Data, Step 2

- Figure 11.5, step 3 of the importing process, prompts the user to specify the type of data in each field. For each field of data, you can stipulate the data as general, text, date, or whether to skip importing a column altogether. When the **Data Preview** screen displays the data in the correct columns, click **Finish**.

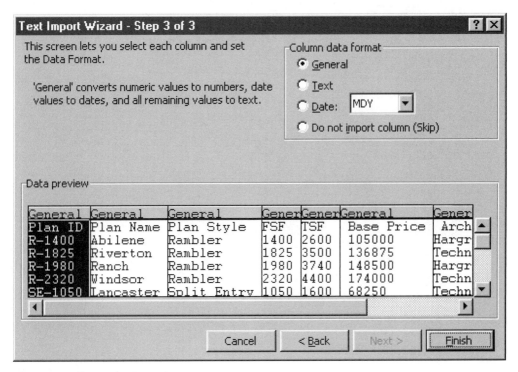

Figure 11.5 Importing Data, Step 3

The import is complete and the Excel file is shown in Figure 11.6 (field names have been bolded for ease of identification). You are now able to manipulate the data, format the various fields, and make any calculations.

	A	B	C	D	E	F	G
1	Plan ID	Plan Name	Plan Style	FSF	TSF	Base Price	Architect
2	R-1400	Abilene	Rambler	1400	2600	105000	Hargreaves
3	R-1825	Riverton	Rambler	1825	3500	136875	Technigraphics
4	R-1980	Ranch	Rambler	1980	3740	148500	Hargreaves
5	R-2320	Windsor	Rambler	2320	4400	174000	Technigraphics
6	SE-1050	Lancaster	Split Entry	1050	1600	68250	Technigraphics
7	SE-1230	Cambridge	Split Entry	1230	1960	79950	Design Specialists
8	SE-1500	Dearborne	Split Entry	1500	2500	97500	Hargreaves
9	SL-1300	Sumpter	Split Level	1300	2100	91000	Design Specialists
10	SL-1454	Olympic	Split Level	1454	2408	101780	Technigraphics
11	SL-1672	Lexington	Split Level	1672	2844	117040	Hargreaves
12	SL-1896	Uintah	Split Level	1896	3292	132720	Technigraphics
13	TS-1850	Vermont	Two-Story	1850	2850	138750	Technigraphics
14	TS-1930	Manchester	Two-Story	1930	2900	144750	Design Specialists
15	TS-2120	Newberry	Two-Story	2120	3250	159000	Technigraphics
16	TS-2340	Charlotte	Two-Story	2340	3320	175500	Hargreaves
17	TS-2560	Richmond	Two-Story	2560	3580	192000	Technigraphics

Figure 11.6 Text File Imported into Excel

Exporting Data

Data from Excel can be exported to other applications and as other data types. Sometimes you will want to manipulate data in Excel and then import it into another application such as an accounting or scheduling software. To export your spreadsheet data for use in another application:

1. Click **File**
2. Click **Save As**, which activates the **Save As** dialog box (Figure 11.7). **Save As** creates a copy of the current Excel file and allows the user to make changes to any or all of the following: file name, disk location, and the file type for the copied file.

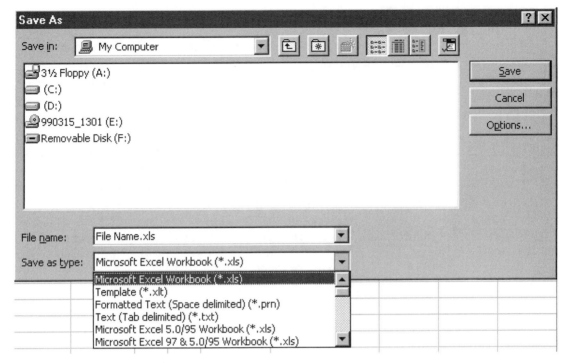

Figure 11.7 The Save As Dialog Box

3. For the new file that will be created, select the path (the location of the file on disk).
4. Enter a new file name.
5. Choose the type of file.

The default selection for the **Save As** type is Microsoft Excel Workbook (*.xls), meaning that the file will be saved in the current version of Excel with a file extention of .xls. File types include current and past versions of Excel, Quattro Pro (wk#, # meaning 1, 2, 3, or some numbered version), and Lotus (wq#). Excel files can also be saved as templates (.xlt), comma-delimited (.csv), space-delimited (.prn), or tab-delimited (.txt) text files, Dbase database file (.dbf), Macintosh (.txt or .csv), O/S2 or MS DOS (.txt or .csv), Data Interchange Format (.dif), Symbolic Link (.slk), or Excel Add-in file (.xla)

Data can easily be arranged in Excel for export into a Personal Digital Assistant (PDA) such as a Palm Pilot® or Newton®, etc. Reorder the Excel columns to match the order of the fields in the PDA program and then perform the import from the PDA software.

Copying Text and Objects

Windows-based software applications allow users to copy text and objects from one program to the other. Use the **Edit/Copy** command to create a copy of the text or image and place it on the clipboard. Edit/Cut will also create a copy to the clipboard but will remove the original. The clipboard is a temporary file that holds the copied text or object until it can be placed in the appropriate location.

Put the cursor in the target (destination) application, the location where the text or object will be copied. Click **Edit/Paste** and the copied text or object will be displayed. Copy and Paste can be done by mouse dragging the text or object *while* holding the **Ctrl** key down. Cut and Paste can be done by mouse dragging the text or object *without* holding the **Ctrl** key down. Arrange the applications you are using so that they are showing side-by-side. Use the minimize button to resize each application window to a smaller size (Figure 11.8).

Figure 11.8 Reducing the Application Window Size

Then resize the window by dragging the sides or corners. Move the windows by clicking on the title bar and dragging to a new location.

Figure 11.9 Resize the Excel Window by Dragging the Edge of the Screen

Once the applications are next to each other, it is a simple task to drag the text or object to the new location from one application to the other (Figure 11.10).

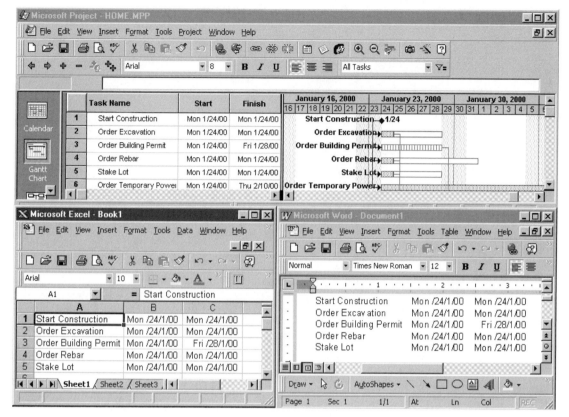

Figure 11.10 Arranging Application Windows

All or part of an Excel worksheet can be copied to another application, to a Microsoft Word® document, for example. Within Word, you now have a functional table. Figure 11.11 shows the plan data (that we created earlier in Excel) after it was pasted into an MS Word document. All of the features that are included in a Word table are accessible.

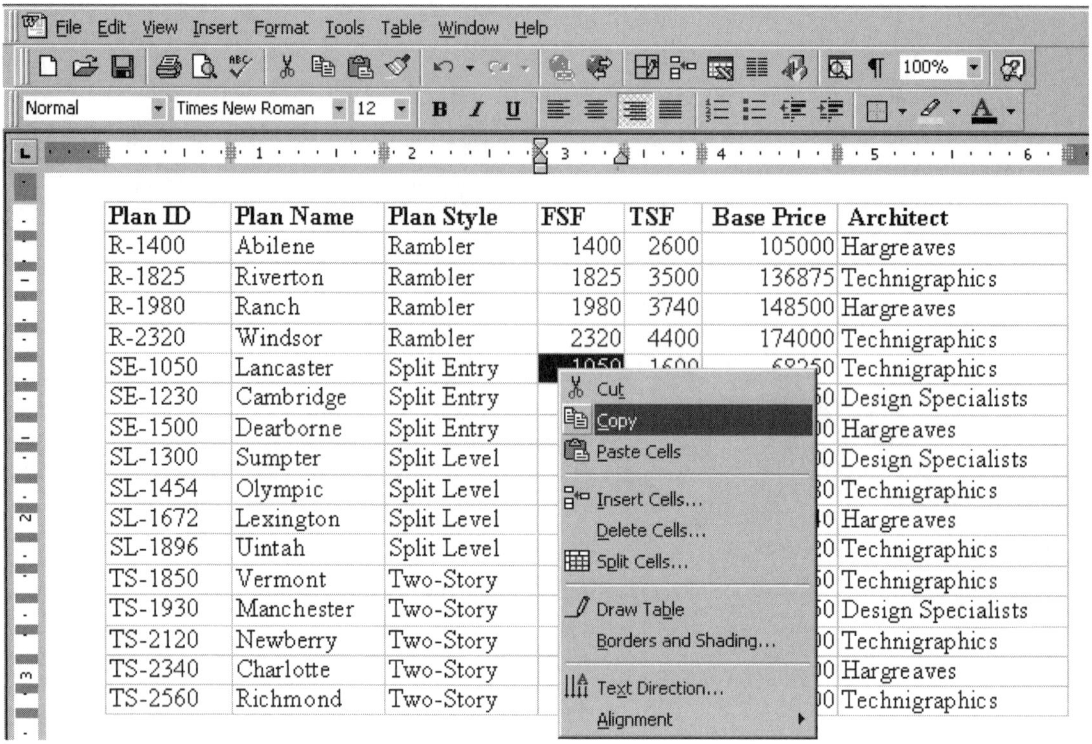

Figure 11.11 Pasting an Excel Worksheet into an MS Word Document

Many times I've seen users create tables in Word to display data when it would be much easier and faster to enter the data into a spreadsheet and copy it into the Word document.

Object Linking and Embedding

There are times when you want text, data, or objects in one file to reflect the changes made to the same text, data, or objects in another file; that is, if you change the data in one file, you would like it to be mirrored in the data of another file. One builder links the construction dates that are generated from his schedule in MS Project to a report in Excel for use by the superintendents. The scheduling information is also linked to a report in MS Word that gets faxed to suppliers and trade subcontractors. The Word report gives the specifications for construction tasks and deliveries as well as the scheduled start and finished dates. Object linking and embedding (OLE) is a technology (not limited to Microsoft products), which allows different applications to communicate with each other and to share data.

Understanding Embedded Objects

The following are characteristics of embedded objects:

- Embedded objects become part of the destination file (the file to which they were copied).
- There is no link between embedded objects and their source files (the files from which they were copied).
- When you edit an embedded object, there is no change to the original (source) worksheet.
- To change an embedded object, double-click the object to open and edit it in the source program.

In MS Word, when you use the **Edit/Paste Special** command, select the **Microsoft Excel Worksheet Object** from the menu, and click on **Paste**, Excel is embedded into the Word document (Figure 11.12). A second way to embed an object is to use the **Insert/Object** command.

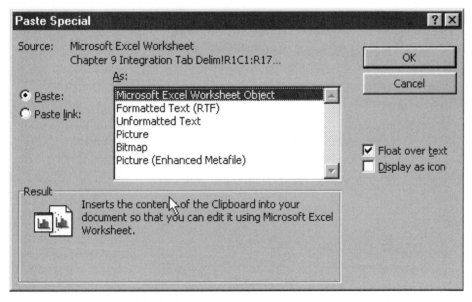

Figure 11.12 Paste Special Dialog Box

An embedded object appears as it would if it were printed from an Excel worksheet (Figure 11.13). The embedded object in Word is not linked to the source file in Excel.

Plan ID	Plan Name	Plan Style	FSF	TSF	Base Price	Architect
R-1400	Abilene	Rambler	1400	2600	105000	Hargreaves
R-1825	Riverton	Rambler	1825	3500	136875	Technigraphics
R-1980	Ranch	Rambler	1980	3740	148500	Hargreaves
R-2320	Windsor	Rambler	2320	4400	174000	Technigraphics
SE-1050	Lancaster	Split Entry	1050	1600	68250	Technigraphics
SE-1230	Cambridge	Split Entry	1230	1960	79950	Design Specialists
SE-1500	Dearborne	Split Entry	1500	2500	97500	Hargreaves
SL-1300	Sumpter	Split Level	1300	2100	91000	Design Specialists
SL-1454	Olympic	Split Level	1454	2408	101780	Technigraphics
SL-1672	Lexington	Split Level	1672	2844	117040	Hargreaves
SL-1896	Uintah	Split Level	1896	3292	132720	Technigraphics
TS-1850	Vermont	Two-Story	1850	2850	138750	Technigraphics
TS-1930	Manchester	Two-Story	1930	2900	144750	Design Specialists
TS-2120	Newberry	Two-Story	2120	3250	159000	Technigraphics
TS-2340	Charlotte	Two-Story	2340	3320	175500	Hargreaves
TS-2560	Richmond	Two-Story	2560	3580	192000	Technigraphics

Figure 11.13 Embedding an Excel Spreadsheet into a Word Document

To edit the embedded object, double click on the object and Excel takes control (Figure 11.14). All of Excel's capabilities are made available to edit the object. When the user clicks off of the object, control is returned to Word.

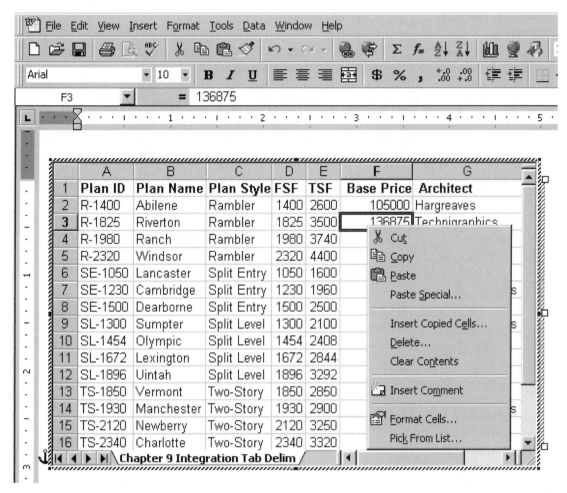

Figure 11.14 In-Place Editing of an Embedded Excel Object in a Word Document

Understanding Linked Objects

Use linked objects if you want the information to reflect any changes to the original data, or if the file size is a consideration. With a linked object, the original information remains stored in the source file. The destination file displays a representation of the linked information but stores only the location of the original data. The linked information is updated automatically if you change the original data in the source file. For example, if you select a range of cells in a Microsoft Excel workbook and then paste the cells as a linked object in a Word document, the information is updated in Word if you change the information in your workbook.

The following are characteristics of linked objects:

- With a linked object, the original information remains stored in the source file.
- The linked object is a marker that is placed in the destination file which points back to the source file.
- When the source data is changed, the linked object is also changed.
- Unlike in-place editing that takes place with embedded objects, if you double click on the linked object (in Word, for example), the source worksheet is loaded into Excel.

To create a linked object, copy an object from the source (it will automatically be placed on the clipboard). In the destination file, use the **Edit/Paste Special** command and choose the **Paste Link** option button (see Figure 11.12—except that the **Paste Link** option button would be activated).

The following simple example demonstrates how different applications can integrate together to create a powerful and useful tool. Each software application performs the functions it does best and then passes the information on to the next software program. The user inputs information only once. The principles shown here can be used in many circumstances and with other software packages.

In this example, information about a home's construction schedule is processed in MS Project. The user inputs each activity including activity durations and activity relationships. Project calculates the Early Start and Early Finish dates for each activity (Figure 11.15).

Integration: Using Excel with Other Programs 179

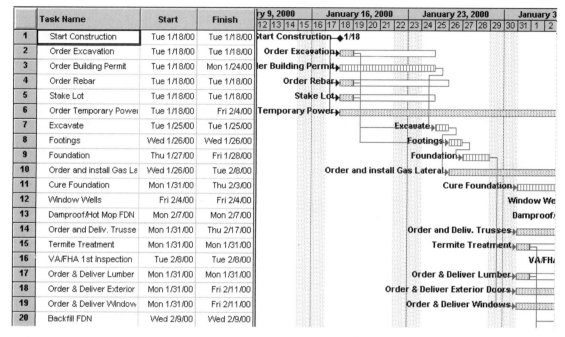

Figure 11.15 Sample MS Project Schedule

Notice the two parts to the MS Project screen: The left half shows a spreadsheet-like representation of the project data, including the task name, early start date, and early finish date. Hidden beneath the Gantt chart (extending to the right of the columnar data) are the late start date, late finish date, free slack, and total slack.

1. Select all of this information from the task name, row one to the total slack, row 83.
2. Copy this range
3. Move to a new workbook in Excel and select cell **A2** on one of the sheets.
4. Click **Edit**.
5. Click **Paste Special**.
6. Select **Paste Link**.
7. Choose the Microsoft Excel Format.

The tasks, dates, and slack will now be linked to the Excel spreadsheet. You will need to enter the headings in cells A1 through G1. Anytime a change is made to the schedule, those changes will be reflected on the worksheet in Excel. For later use, name the range of cells, A2 through G84, ActivityDB.

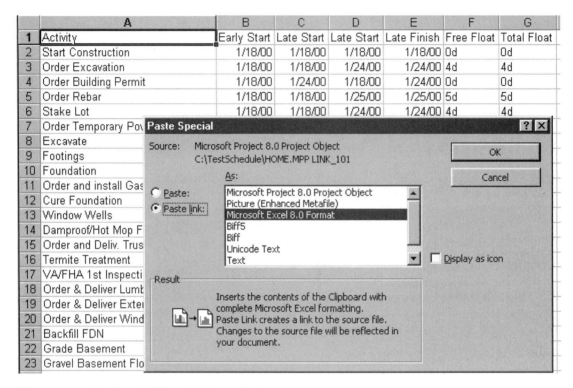

Figure 11.16 Pasting a Link from Project to Excel

Suppose you wanted to be able to choose a specific activity and have Excel return the Early Start and Early Finish dates.

On a separate worksheet in Excel, create a Data Validation in cell A2 (Figure 11.17) that will return values the list of scheduled activities (tasks) on the sheet with the linked information from MS Project.

Figure 11.17 Looking up Information

In cells B2 and C2 create VLOOKUP formulas that will look up the Early Start and Early Finish dates for the activity selected in cell A2.

The VLOOKUP formula for B2 could be written:

=IF(A2="","",VLOOKUP(A2,ActivityDB,2,FALSE))

The formula reads: If **A2** is blank then **B2** will be remain blank, otherwise, the formula looks at the value in **A2** and matches it with a value in the first column of **ActivityDB**. The Early Start date (in the second column), which corresponds to the activity that was chosen in **A2**, will be returned.

Use a similar **VLOOKUP** formula for **C2**:

=IF(A2="","",VLOOKUP(A2,ActivityDB,3,FALSE))

Use 3 for the **Column_Index_Number** instead of 2.

Suppose also that you had a contract document in MS Word that had a statement indicating when a task should be started. Copy cell **A2** in Excel and activate MS Word by clicking on the Word tab at the bottom of the page or by pressing **Alt+Tab**. Place the cursor where you want the task information inserted and click **Edit/Paste Special** and choose **Paste Link**. Repeat the process for **B2** on the Excel sheet. The statement in your Word document may look something like Figure 11.18.

Figure 11.18 Linked Word Document

As planning for the project progressed, it was necessary to move the start date of the project to the twenty-fourth of January. Return to MS Project, place the cursor over the Start Construction milestone on the Gantt chart (Figure 11.19) and drag the start date to January 24. Instantly, the project's schedule dates are recalculated and the bar chart adjusts for the new start date.

182 Estimating with Microsoft Excel

	Task Name	Start	Finish
1	Start Construction	Mon 1/24/00	Mon 1/24/00
2	Order Excavation	Mon 1/24/00	Mon 1/24/00
3	Order Building Permit	Mon 1/24/00	Fri 1/28/00
4	Order Rebar	Mon 1/24/00	Mon 1/24/00
5	Stake Lot	Mon 1/24/00	Mon 1/24/00
6	Order Temporary Power	Mon 1/24/00	Thu 2/10/00
7	Excavate	Mon 1/31/00	Mon 1/31/00
8	Footings	Tue 2/1/00	Tue 2/1/00
9	Foundation	Wed 2/2/00	Thu 2/3/00
10	Order and install Gas La	Tue 2/1/00	Mon 2/14/00
11	Cure Foundation	Fri 2/4/00	Wed 2/9/00

Figure 11.19 Making a Change in the Start Date of Construction

Activate the Excel worksheet that has the List Validation and VLOOKUP formulas and verify that the start and finish dates reflect the change made to the project schedule (Figure 11.20).

B2 =IF(A2="","",VLOOKUP(A2,ActivityDB,2,FALSE))

	A	B	C
1	Schedule Activity	Start	Finish
2	Start Construction	January 24, 2000	January 24, 2000
3			

Figure 11.20 Changes in the Schedule are Automatically Reflected in Excel

Activate MS Word and observe that the changes are also reflected in the contract document (Figure 11.21).

Please begin the task: <u>Start Construction</u> on <u>January 24, 2000</u>!

Figure 11.21 Changes to the Schedule are Instantly Reflected in Word

Move again to Excel and choose Footings. The Early Start and Finish dates for the footings are displayed (Figure 11.22). They correspond to the Footings' early start and finish dates in the MS Project schedule.

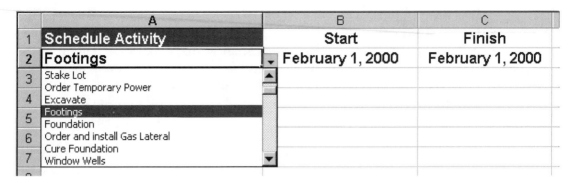

Figure 11.22 Select a Different Activity: Excel Returns the Early Start and Finish Dates

Now verify that the new selection (Footings) and its start date are updated on the Word document. Activate Word and you should see the sentence as shown in Figure 11.23. (The linked data has been underlined for distinction.)

 Please begin the task: <u>Footings</u> on <u>February 1, 2000</u>!

Figure 11.23 Verifying the Automatic Updates to the MS Word Document

Conclusion

This has been a small example to show how programs can integrate, but the possibilities are endless. Data from many different accounting packages can be exported to text files, manipulated in Excel, reported in Word and can, in many cases, be imported back into the accounting program for additional processing. Customer information such as color selections and construction options can be processed in Excel and linked to contract documents in Word. Information from databases in MS Access can be linked to documents in Word or spreadsheets in Excel. Demonstrations were made primarily with Microsoft products, but any software that supports Object Linking and Embedding can exchange data, and many programs that do not support OLE can export and import information between software packages.

We have seen many changes in building over the past few decades, and especially since the advent of personal computers. Some builders and remodelers have resisted changes in technology, but gradually, more and more builders are learning software and using computers to make their work more efficient and effective. The problem with most builders is that they feel like they're just too busy to take time out to learn new technology. Remember, you will be more successful if you regularly make time to sharpen your saw.

Computers are here to stay and will be an ever increasing part of doing business. The time you invest now to learn how to use them more effectively, will pay great dividends in the future.

Appendix: How to Use the CD

This book assumes you know how to estimate and that you have tackled some computer estimating. You will learn more and faster if, as you read this book, you try these ideas and concepts on your Excel spreadsheet. Take a chance to explore on your own and discover new ways to make your spreadsheets more efficient and effective. To make that easier, a CD is included with this book. It contains the following files:

Chapter 1	Principal.xls
Chapter 2	Database.xls
Chapter 3	Cost Breakdown Summary.xls
Chapter 4	Start.xls
Chapter 4	Finish.xls
Chapter 5	Linking Start.xls.
Chapter 4	Linking Finish.xls
Chapter 6	Formulas Start.xls
Chapter 6	Formulas Finish.xls
Chapter 7	Interest & Profit.xls
Chapter 8	Macros.xls
Chapter 9	Option Buttons.xls
Chapter 10	UserForms Example 1.xls
Chapter 10	UserForms Example 2.xls
Chapter 11	Embedded Excel Object.doc
Chapter 11	Excel Table Copy_Paste.doc
Chapter 11	Integration Comma Delimited.doc
Chapter 11	Integration Comma Delimited.txt
Chapter 11	Integration PlanDB AutoFilter.xls
Chapter 11	Integration Tab-Delimited File.xls

HOME.MPP
Schedule Activities.xls
Schedule letter.doc
Estimator*PRO* Sample

For best results, these files should be installed to the **C:\NAHB\Estimating with MS Excel** path. This is particularly important for these files that go with Chapter 11:

- HOME.MPP
- Schedule Activities.xls
- Schedule letter.doc

Note that when opening some of these files a box may come onto your screen that states:

- The workbook you are opening contains Macros.
- If you are sure this workbook is from a trusted source, click Enable Macros.
- If you are not sure and want to prevent any macros from running, click Disable Macros.

All files have been tested for viruses and will not harm your computer. Click **Enable Macros** in order to make full use of the files.

If you open the sample **EstimatorPRO.xlt** file and get a message stating in effect "Excel cannot calculate a formula. Cell references in the formula refer to . . . a circular reference." Click **OK**. Then **Tools/Options/Calculation**. Then click on the box next to **Iteration** and click **OK**.

Also, you will notice some files are "start" and some are "finish." In this way, you can see how the completed Excel file should look—just in case you get stuck. However, the lessons are laid out in such a way that even if you're not terribly familiar with Excel, you'll be unlocking its power in no time.

Resources

Blattner P. and Ulrich L. *Using Microsoft Excel 2000, Special Edition*. QUE. Indianapolis, IN., 1999.

Chester, T. and Alden, R. *Mastering Excel 97 Fourth Edition*. Sybex. San Francisco, CA., 1997.

Christofferson, Jay. Survey of attendees of the "Estimating for Builders: Secrets to Unlocking the Power of Excel." Seminar presented at the International Builders' Show. Dallas, TX., January, 1999.

Cottingham, Marion. *Excel 2000 Developer's Handbook*. SYBEX, San Francisco, CA., 1999.

Dodge, M. and Stinson. C. *Running Microsoft Excel 2000*, Microsoft Press. Redmond, WA., 1999.

Green, John. *Excel 2000 VBA Programmer's Reference*. Wrox Press Ltd. Arden House, Birmingham, AL, 1999.

Harris, Matthew. Excel *2000 Programming in 21 Days*. SAMS, Indianapolis, IN., 2000

―――― *Teach Yourself Microsoft Excel 2000 Programming in 21 Days*. Sams Publishing. Indianapolis, IN.,1999.

Reisner, Trudy. *Teach Yourself Microsoft Excel 2000 in 24 Hours*. Sams Publishing. Indianapolis, IN., 1999.

Stewart, Laura. *Using Microsoft Office 2000, Platinum Edition*. QUE. Indianapolis, IN., 1999.

Taylor, Dennis. *Teach Yourself Microsoft 2000*. IDG Books Worldwide, Inc. Foster City, CA., 1999.

Walkenbach, J. and Maguiness, D. *Excel 5 for Windows Handbook*, 2nd Edition. IDG Books Worldwide, Inc. Foster City, CA., 1994.

Walkenbach, John. *Microsoft Excel 2000 Bible.* IDG Books Worldwide, Inc., Foster City, CA., 2000.

Walkenbach, John. *Power Programming with VBA.* IDG Books Worldwide, Inc., Foster City, CA., 2000.

Wempen F. Payne. D. *The Essential Excel 2000 Book.* Prima Tech. Rocklin, CA., 1999.

Wyatt, Allen. *Hands on Excel 2000.* Jamsa Press, Houston, TX., 1999.

Index

$ (dollar sign), 10, 37–38, 64, 142
= (equal sign), 31
#N/A (error sign), 60–61, 77–78
% (percent sign), 10
^ (power sign), 96

A

Addition, 31, 32–33, 34–35, 46–47
AutoFill feature, 15–16, 33–34
AutoSum function, 32–33, 34–35, 46–47

B

Bold font, 16–17
Builder margin
    calculating, 108–10
    company overhead costs and, 108, 109
    direct costs and, 107–9
    sales price and, 108–10
Building Standards, 86
Building Valuation Data, 86
Buttons. *See* Command buttons

C

CD, for this book, 185–86
Cells. *See also* Named cells
    address of, 5
    anchoring references to, 142
    blocks of, 7
    Check Boxes in, 141–42
    comments in, 93–94
    copying, 17
    entering data in, 15–16
    formatting, 10, 16–17
    moving, 17
    non-contiguous, 8
    protecting, 89–90
    selecting, 7–8
    viewing formulas in, 62–63
Check Boxes, 140–42
    features of, 140
    IF Function for, 141–42
    linking cells to, 141–42
    setting up, 141
Circular references, 110–12
    calculating sales price and, 110–12
    defined, 110
    iteration and, 111
    toolbar, 111–12
    tracing accidental, 111–12
Col_index_number, 51
Color, formatting, 16–17, 139, 146
Colored font, 16–17
Columns, 5
    adding sums in, 34–35
    naming, 8–9
    Transpose function and, 42
    in VLOOKUP function, 51
    width, 18
Combo Boxes, 147
Comma-delimited files, 165–66
Command buttons, 119–22
    assigning macros to, 119–20

Command buttons (*Continued*)
   code for, 158–62
   creating custom, 121–22
   editing, 120–21
   in UserForms, 155–56, 158–62
Comments, cell, 93–94
Company overhead costs (G&A), 108, 109
Concrete estimates, 72–83
   calculated quantities in, 72, 73–74
   cubic yards (CY) and, 72–73
   custom formatting for, 84–85
   List Validation for, 81–83
   MATCH function and, 78–81
   rebar calculations and, 83–85
   rounding numbers in, 75–76
   square feet (SF) and, 73
   take-off quantities, 72–73
   VLOOKUP function and, 76–78
   waste factoring and, 74–75
Connection fees, 91–92
Constants, 12–14
Construction loan interest, 103–7
   construction S-curve and, 105–6
   costs, 103–4
   origination fees, 104
   reserve calculation, 105–7
Construction valuation, 86–89
Copying
   cell contents, 17
   formulas, 10–12, 33–35, 38–44
   objects, 172–74
   text, 172–74
Cost breakdown summary sheets, 19–22, 29–44
   advantages of, 29–30
   CD files for, 28, 44, 59–60
   cost control and, 21–22
   cost variance and, 21
   creating custom, 19, 28
   detail sheets. *See* Detail sheets
   direct costs in, 20
   formulas. *See* Formulas
   layout of, 30–31
   typical contents, 20–21
   uses of, 19
   using, 29–31
Costs
   actual vs. estimated, 29
   builder margin and, 107–10
   company overhead, 108, 109

   construction loan, 103–4
   direct, 20, 107–9
   extending, 46–47, 77
   sales price from, 108–10
Cost variance, 21
Cubic yards (CY)
   calculating concrete in, 73–74
   concrete waste in, 74–75
   defined, 73
Currency, formatting, 10, 16–17, 37–38, 64
Cursor
   changing settings of, 7
   moving, 6–7
   recording macros and, 116
Custom formatting, 84–85
CY. *See* Cubic yards (CY)

D

Data. *See also* Named cells
   AutoFill feature, 15–16, 33–34
   entering, 15–16
   exporting, 170–71
   finding, in tables, 14–15
   importing, 167–70
   looking up. *See* VLOOKUP function
   saving, as other file types, 170–71
   sorting, 42–44
   transposing, 42
Databases, 26–28
   advantages of, 28
   comma-delimited, 165–66
   naming, 48–49
   tab-delimited, 166
Data Validation, 53–57
   advantages of, 53, 57
   criteria, 54
   dialog box, 54, 55
   drop-down lists, 54–56, 63
   linked objects and, 180–83
   List Validation, 81–83
   source lists, 54–56
   time savings with, 53, 57
   UserForms and, 149–51
   VLOOKUP function and, 57, 150–51, 180–83
Dates
   current, 35–36
   formatting, 35–37
   Now function and, 35–36
Decimal places, formatting, 38

Detail sheets, 22–26, 45–58
  CD files for, 58, 60
  cost estimates on, 22
  databases and, 26–28
  data validation on, 53–57, 149–51, 180–83
  extending costs on, 46–47, 77
  format, 26, 45–47
  formulas. *See* Formulas; VLOOKUP function
  recalculations with, 24
  what-if scenarios, 23–25
Direct costs
  builder margin and, 107–10
  cost breakdown summary sheets and, 20
  defined, 107–8
  profit margin and, 107–9
Division, 31
Dollar signs ($), 10, 37–38, 64, 142
Drawing toolbar, 4
Drop-down lists, 54–56, 63

E

Embedded objects. *See* Object linking and embedding (OLE)
Equal sign (=), 31
Error sign (#N/A), 60–61, 77–78
Excel. *See* Microsoft Excel
Exponentiation, 31
Exporting data, 170–71
Extending costs, 46–47, 77

F

Fees
  connection, 91–92
  impact, 91–92
  permit, 86–89
Finding information, 14–15
Font, formatting, 16–17
Formatting
  cells, 10, 16–17
  color, 16–17, 139, 146
  currency, 10, 16–17, 37–38, 64
  custom, 84–85
  dates, 35–37
  decimal places, 38
  font, 16–17
  List Boxes, 145–46
  Now function, 35–36
  percentages, 10, 16–17
  text, 16–17, 139, 146

toolbars, 4, 17
Form tools. *See also* UserForms
  Check Boxes, 140–42
  Combo Boxes, 147
  List Boxes, 144–47
  Option Buttons, 136–40
  Scroll Bars, 142–43
  Spinners, 143–44
  toolbar for, 119, 135–36
Formulas, 31–38
  anchoring cell references in, 142
  AutoFill feature, 15–16, 33–34
  changing, 12
  circular references in, 110–12
  constant names in, 12–14
  copying, 10–12, 33–35, 38–44
  equal sign (=) and, 31
  Function Wizard (fx) for, 35, 49–52
  If, Then, Else, 60–61, 70
  looking up data in. *See* VLOOKUP function
  named cells in, 9, 11–14
  naming, 10–12
  nesting, 61, 81, 89
  operators for, 31, 38–44
  Paste Special command and, 38–44
  starting, 31
  variables in, 11, 90
  viewing, 62–63
  writing, 31–33, 52
Functions. *See also* IF Function; VLOOKUP function
  advantages of, 47, 57, 71
  AutoSum function, 32–33, 34–35, 46–47
  MATCH function, 78–81, 91–92, 98, 99
  Now function, 35–36
  ROUNDUP function, 75–76, 83, 84, 87, 97–98
  Square root (SQRT) function, 96
  Transpose function, 42
Function Wizard (fx), 35, 49–52

G

General and Administrative. *See* Company overhead costs (G&A)

H

Hardware requirements, 3
Header rows, 43
Help, 3
Horizontal, transposing, 42

Hyperlinks, 65–70. *See also* Linking; Object linking and embedding (OLE)
  advantages of, 68
  creating, 67–68
  defined, 65
  inserting, 65–66
  to Internet files, 69–70
  naming destinations of, 66
  to other files, 68–69

I

Icons, custom
  changing images for, 125–28
  creating toolbar for, 123–25
IF Function, 60–61
  #N/A error signs and, 60–61, 77–78
  Check Boxes and, 141–42
  defined, 61
  nesting formulas, 61, 89
  Option Buttons and, 139–40
  VLOOKUP function and, 60–61
Impact fees, 91–92
Importing data, 167–70
Inserting
  comments, 93–94
  hyperlinks, 65–66
  worksheets, 5, 6
Italics, 16–17
Iteration, 111

L

Linking, 59–60, 63–70. *See also* Hyperlinks; Object linking and embedding (OLE)
  advantages of, 59
  CD files for, 59–60
  methods, 63–64
  multiple programs, 178–83
  objects, 178–83
  Paste Special for, 64, 178, 179
List Boxes, 144–47
  code for, 158–62
  defined, 144
  formatting, 145–46, 154–55
  RowSource and, 156–57
  in UserForms, 154–55, 156–62
  VLOOKUP function for, 146–47
Lists, drop-down, 54–56, 63
List Validation, 81–83
Loan-to-value ratio (LTV), 104

Looking up data. *See* VLOOKUP function
Lookup_array, 79, 80
Lookup_value, 50, 79
LTV. *See* Loan-to-value ratio (LTV)

M

Macros
  advantages of, 113–14, 134
  attaching, to objects, 119–20, 122–23
  command buttons for, 119–22
  cursor position and, 116
  custom icons for, 123, 125–28
  custom menu items for, 128–31
  custom toolbars for, 123–25
  defined, 113
  editing, 132–34
  recording, 114–17
  relative references in, 115
  running, 117–20
  storing, 115
  UserForm, 151–54
  Visual Basic for Applications, 113, 132–34, 152–54
Match_type, 79
MATCH function, 78–81, 91–92, 98, 99
Menu toolbar
  custom items on, 128–31
  location of, 4
  underlined items on, 130
Microsoft Excel
  comma-delimited files and, 165–66
  copying text/objects, 172–74
  exporting data, 170–71
  help, 3
  importing data, 167–70
  integrating, with other programs, 165–83
  saving, as other file types, 170–71
  tab-delimited files and, 166
  version requirements, 3
Microsoft Project, 178–80, 181
Microsoft Word, 173–74, 175–77, 182–83
Mistakes, undoing, 3–4
Moving
  cells, 17
  cursor, 6–7
  Option Buttons, 137
  worksheets, 6
Multiplication, 31, 40–41

N

Named cells
  constants as, 12–14
  finding information with, 14–15
  in formulas, 9, 11–14
  formulas as, 10–12
  naming, 8–9
Nesting formulas
  IF Functions, 61, 89
  MATCH, VLOOKUP functions, 81
  maximum number of, 89
Now function, 35–36

O

Object linking and embedding (OLE), 174–83. *See also* Hyperlinks; Linking
  defined, 174
  embedding objects, 175–77
  linked objects, 178–83
  with other programs, 175–83
Operators, 31, 40–41
Option Buttons, 136–40
  activating, 139–40
  defined, 136
  Group Boxes for, 139, 140
  IF Function for, 139–40
  labeling, 138–39
  moving, 137
  referencing, to cells, 137–38
  resizing, 137
  setting up, 136–37
Origination fees, 104

P

Paste Link command, 64, 178, 179
Paste Name command, 51
Paste Special command, 38–44
  dialog box, 39
  embedding objects with, 175–77
  linking with, 64, 178, 179
  operators, 40–42
  Transpose function, 42
PDA, 171
Percentage(s)
  calculating, 31
  formatting, 10, 16–17
  sign (%), 10
Permit fees, 86–89, 90
Personal Digital Assistant (PDA), 171

Personal Macro Workbook, 115, 118–19
Power sign (^), 96
Profit
  builder margin and, 107–10
  company overhead costs and, 108, 109
  defined, 108
  direct costs and, 107–9
  sales price and, 108–10
Protecting cells, 89–90
Pythagorean formula, 96

R

Range_lookup, 51–52
Rebar calculations, 83–85
Repair shop phenomenon, 25–26
Roofing calculations, 94–101
  CD files for, 101
  labor rate, 98–99
  MATCH function and, 98, 99
  Pythagorean formula, 96
  ROUNDUP function and, 97–98
  shingle quantity, 96–98
  slope factor, 96
  starter strips, 98
  sub-bid methods, 99–101
  VLOOKUP function and, 98, 99
  what-if scenarios, 23–25, 101
Rounding numbers, 75–76, 87
ROUNDUP function, 75–76, 83, 84, 87, 97–98
Rows, 5
  height, 18
  naming, 8–9
  Transpose function and, 42
RowSource, 156–57

S

Sales price
  builder margin and, 107–10
  calculating, 108–10
  circular references and, 110–12
  company overhead costs and, 108, 109
  direct costs and, 107–9
Scroll Bars
  creating, 142–43
  features of, 142
  Spinners and, 143–44
S-curve, construction, 105–6
SF. *See* Square feet (SF)
Slope factor, 96
Sorting data, 42–44

Spinners, 143–44
Spreadsheets
   advantages of, xix–xx, 23–26, 28
   enhancing. See Form tools
   formatting. See Formatting
   learning, value of, 1–2
   repair shop phenomenon and, 25–26
   what-if scenarios and, 23–25, 101
Square feet (SF), 73
Square root (SQRT) function, 96
Standard toolbar, 4
Status toolbar, 4
Stop Recording toolbar, 115, 116–17
Storing macros, 115
Subtraction, 31, 32
SUM function. See AutoSum function

T

Tab-delimited files, 166
Table_array, 51
Take-off quantities, 72–73
Text, formatting, 16–17, 139, 146
Text Import Wizard, 167–69
Toolbars, 4–5
   Circular Reference, 111–12
   creating custom, 123–25
   custom icons for, 123–28
   Drawing, 4
   Formatting, 4, 17
   Forms, 119, 135–36
   Menu, 4, 128–31
   Standard, 4
   Status, 4
   Stop Recording, 115, 116–17
Transpose function, 42

U

Underlined font, 16–17
Undoing mistakes, 3–4
UserForms
   assigning, 156–57
   closing, 159–60
   code for, 158–62
   Command buttons in, 155–56, 158–62
   creating, 151–54
   data validation with, 149–51
   double click event in, 160–61
   features of, 149, 150–51
   List Boxes in, 154–55, 156–62
   macros for, 151–54

   RowSource and, 156–57
   Visual Basic Editor for, 152–54
   VLOOKUP function and, 150–51

V

Validation. See Data Validation
Valuation, construction, 86–89
Variables
   naming, 11
   working with, 90
Vertical, transposing, 42
Visual Basic Editor, 133–34, 152–54
Visual Basic for Applications (VBA), 113, 132–34
VLOOKUP function, 47–53
   advantages of, 47, 57
   Col_index_number and, 51
   for Concrete estimates, 76–78
   in connection/impact fee calculations, 91–92
   Data Validation and, 57, 150–51, 180–83
   defined, 47
   dialog box, 50
   Function Wizard (fx) for, 49–52
   IF Function and, 60–61
   List Boxes and, 146–47
   Lookup_value and, 50
   MATCH function and, 78–81
   Range_lookup and, 51–52
   roofing labor calculation, 98, 99
   Table_array and, 51
   UserForms and, 150–51

W

What-if scenarios, 23–25, 101
Wizards
   Function (fx), 35, 49–52
   Text Import, 167–69
Workbooks
   storing macros in, 115
   using, 4–8
Worksheets. See also Cells
   adding (inserting), 5, 6
   AutoFill feature, 15–16, 33–34
   defined, 5
   deleting, 5
   moving, 6
   moving cursor in, 6–7
   naming, 6
   naming cells in, 8–9
   selecting cells in, 7–8